Introduction to Paleoanthropology

David Speakman

Originally published on Wikibooks.org
Wikibooks is a project of the Wikimedia Foundation
THIS PUBLICATION IS NOT OFFICIALY
SPONSORED BY WIKIBOOKS OR WIKIPEDIA

Published by
Seven Treasures Publications

Published by
Seven Treasures Publications
SevenTreasuresPublications@gmail.com
Fax 413-653-8797

Available for purchase from www.Amazon.com and www.BN.com
For a full list of available wiki titles search for
"seven treasures publications"

Printed in the United States of America

ISBN 978-0-9800707-5-0

Introduction to Paleoanthropology

Edition 1.0 24th February 2006

From Wikibooks, the open-content textbooks collection

Note: current version of this book can be found at
http://en.wikibooks.org/wiki/Introduction_to_Paleoanthropology

Authors:

David Speakman (Wikibooks user: Davodd) and anonymous authors.

WHAT IS ANTHROPOLOGY?

The study of anthropology falls into four main fields:

1.   Sociocultural anthropology

2.   Linguistics

3.   Archaeology

4.   Physical anthropology

Although these disciplines are separate, they share common goals. All forms of anthropology focus on the following:

- Diversity of human cultures observed in past and present.

- Many scientific disciplines involved in study of human cultures.

- Examples include: Psychology, biology, history, geography among others.

- Anthropology holds a very central position in the world of science.

- There is a long academic tradition in modern anthropology which is divided into four fields, as defined by Franz Boas (1858-1942), who is generally considered the father of the field.

Sociocultural anthropology/ethnology

This field can trace its roots to global colonial times, when the European and American dominance of overseas territories offered scholars access to different cultures. Over the years, this field has expanded into urban studies, gender studies, ethnic studies and medical anthropology.

Linguistics

This study of human speech and languages includes their structure, origins and diversity. It focuses on comparison between contemporary languages, identification of language families and past relationships between human groups. It looks at:

- Relationship between language and culture

- Use of language in perception of various cultural and natural phenomena

- Process of language acquisition, a phenomenon that is uniquely human, as well as the cognitive, cultural, and biological aspects involved in the process.

- Through linguistics we can trace the migratiion trails of large groups of people (be it initiated by choice, by natural disasters, by social and polical pressures). In reverse, we can trace movement

3

and establish the impact of the political, social and physical pressures, by looking at where and when the changes in linguistic usage occured. --65.135.47.190 02:20, 29 Dec 2004 (UTC)

Archaeology

Is the study of past cultures. It uses very specific study methods, because of limitations of this subfield. It should be noted that recovery and analysis of material remains is only one window to reconstruct past human societies and behaviors. Examples include economic systems, religious beliefs, and social and political organization. Archaeological studies are based on:

- Specific excavation techniques, stratigraphy, chronology

- Animal bones, plant remains, human bones, stone tools, pottery, structures (architecture, pits, hearths).

Physical anthropology

Is the study of human remains within the framework of evolution, with a strong emphasis on the interaction between biology and culture. Physical anthropology has three basic subfields:

- Paleoanthropology

- Osteometry/osteology

- Forensic anthropology

Paleoanthropology

As a subset of physical anthropology, this field relies on the following:

- Research Design: Understanding Human Evolution

Evolution of hominoids from other primates starting around 8 million to 6 million years ago

- Importance of physical anthropology

Evidence of hominoid activity between 8 and 2.5 million years ago usually only consists of bone remains available for study. Because of this very incomplete picture of the time period from the fossil record, various aspects of physical anthropology (osteometry, evolutionary framework) are essential to explain evolution during these first millions of years. Evolution during this time is considered as the result of natural forces only.

- Importance of related disciplines

Paleoanthropologists need to be well-versed in other scientific disciplines and methods, including ecology, biology, genetics and primatology. Through several million years of evolution, humans eventually became a unique species. This process is similar to the evolution of other animals that are adapted to specific environments or "ecological niches". Animals adapted

4

to niches usually play a specialized part in their ecosystem and rely on a specialized diet.

Humans are different in many ways from other animals. Since 2.5 million years ago, several breakthroughs have occurred in human evolution, including dietary habits, technological aptitude, and economic revolutions. Humans also showed signs of early migration to new ecological niches and developed new subsistence activities based on new stone tool technologies and the use of fire. Because of this, the concept of an ecological niche does not apply to humans anymore.

Summary

The following topics were covered:

* Introduced field of physical anthropology;

* Physical anthropology: study of human biology, nonhuman primates, and hominid fossil record;

* Placed paleoanthropology within overall context of anthropological studies (along with cultural anthropology, linguistics, and archaeology);

Further modules in this series will focus on physical anthropology and be oriented toward understanding of the natural and cultural factors involved in the evolution of the first hominids.

INTRODUCTION TO THE LOWER PALEOLITHIC

History of Research

Beginning of the 20th Century

In 1891, Eugene Dubois discovers remains of hominid fossils (which he will call Pithecanthropus) on the Island of Java, South-East Asia. The two main consequences of this discovery:

- stimulates research for "missing link" of our origins

- orients research interest toward SE Asia as possible cradle of humanity

Yet, in South Africa, 1924, discovery by accident of remains of child (at Taung) during exploitation of a quarry. Raymond Dart identifies remains of this child and publishes them in 1925 as a new species - *Australopithecus africanus* (which means "African southern ape"). Dart, a British-trained anatomist, was appointed in 1922 professor of anatomy at the University of the Witwatersrand in Johannesburg, South Africa. This discovery:

- documented the ancient age of hominids in Africa

- questioned the SE Asian origin of hominids, arguing for a possible African origin.

Nevertheless, his ideas not accepted by scientific community at the time:

- major discoveries carried out in Europe (Gibraltar, Germany - Neanderthal, etc.) and Asia (Java)

- remains of this species were the only ones found and did not seem to fit in phylogenetic tree of our origins

- finally considered simply as a fossil ape

It took almost 20 years before Dart's ideas could be accepted, due notable to new discoveries:

- in 1938, identification of second species of Australopithecine, also in South Africa: Paranthropus (Australopithecus) robustus. Robert Broom collected at Kromdraai Cave remains of skull and teeth

- in 1947, other remains of A. africanus found at Sterkfontein and Makapansgat

- in 1948, other remains of P. robustus found at Swartkrans, also by R. Broom

1950s - 1970s

During first half of 20th century, most of discoveries essential for paleoanthropology and human evolution done in South Africa.

After World War II, research centers in East Africa with the couple Mary and Louis Leakey. They discovered major site of Olduvai (Tanzania):

- many seasons of excavations at this site - discovery of many layers (called Beds), with essential collection of faunal remains and stone tools, and several hominid species identified for the first time there;

- In 1959, discovery in Bed I of hominid remains (OH5), named Zinjanthropus (Australopithecus) boisei;

- L. Leakey first considered this hominid as the author of stone tools, until he found (in 1964) in same Bed I other hominid fossils, which he attributed to different species - Homo habilis (OH7).

Another major discovery of a paleoanthropological interest comes from the Omo Valley in Ethiopia:

- from 1967 to 1976, 9 field seasons carried out;

- In 1967, discovery of hominid fossils attributed to new species - *Australopithecus aethiopicus*

- 217 specimens of hominid fossils attributed to five hominid species: *A. afarensis, A. aethiopicus, A. boisei, H. rudolfensis, H. erectus*, dated to between 3.3 and 1 Myrs ago.

Also in 1967, RICHARD LEAKEY starts survey and excavation on east shore of Lake Turkana (Kenya), at a location called Koobi Fora:

- research carried out between 1967 and 1975

- very rich collection of fossils identified, attributed to *A. afarensis* and *A. boisei*

In 1972, a French-American expedition led by Donald Johanson and Yves Coppens focuses on a new locality (Hadar region) in the Awash Valley (Ethiopia):

- research carried out between 1972-1976

- in 1973, discovery of most complete skeleton to date, named Lucy, attributed (in 1978 only) to *A. afarensis*

- more than 300 hominid individuals were recovered

- discoveries allow for detailed analysis of locomotion and bipedalism among early hominids

From 1976 to 1979, MARY LEAKEY carries out research at site of Laetoli, in Tanzania:

- In 1976, she discovers animal footprints preserved in tuff (volcanic ash), dated to 3.7 Myrs ago

- In 1978-1979, discovery of site with three series of hominid (australopithecines) footprints, confirming evidence of bipedalism.

7

1980 - The Present

South Africa

Four australopithecine foot bones dated at around 3.5 million years were found at Sterkfontein in 1994 by Ronald Clarke:

- oldest hominid fossils yet found in South Africa

- They seem to be adapted to bipedalism, but have an intriguing mixture of ape and human features

Since then, eight more foot and leg bones have been found from the same individual, who has been nicknamed "Little Foot".

Eastern Africa

Recent discovery of new *A. boisei* skull is:

- one of the most complete known, and the first known with an associated cranium and lower jaw;

- It also has a surprising amount of variability from other *A. boisei* skulls, which may have implications for how hominid fossils are classified.

Recent research suggests that the some australopithecines were capable of a precision grip, like that of humans but unlike apes, which would have meant they were capable of making stone tools.

The oldest known stone tools have been found in Ethiopia in sediments dated at between 2.5 million and 2.6 million years old. The makers are unknown, but may be either early *Homo* or *A. garhi*

main question is, how have these spieces come to exist in the geographical areas so far apart from one another

Chad

A partial jaw found in Chad (Central Africa) greatly extends the geographical range in which australopithecines are known to have lived. The specimen (nicknamed Abel) has been attributed to a new species - *Australopithecus bahrelghazali*.

In June 2002, publication of major discovery of earliest hominid known: *Sahelanthropus tchadensis* (nickname: "Toumai").

BONE TERMINOLOGY AND THE DEFINING OF HUMANS

Bone Identification and Terminology

Skull

Cranium: The skull minus the lower jaw bone.

Brow, Supraorbital Ridges: Boney protrusions above eye sockets.

Endocranial Volume: The volume of a skull's brain cavity.

Foramen Magnum: The hole in the skull through which the spinal cord passes.

- In apes, it is towards the back of the skull, because of their quadrupedal posture

- In humans it is at the bottom of the skull, because the head of bipeds is balanced on top of a vertical column.

Sagittal Crest: A bony ridge that runs along the center line of the skull to which chewing muscles attach.

Subnasal Prognathism: Occurs when front of the face below the nose is pushed out.

Temporalis Muscles: The muscles that close the jaw.

Teeth

Canines, Molars: Teeth size can help define species.

- Gorillas eat lots of foliage; therefore they are chewing all day and have great canines

- Humans are omnivorous and have small, more generalized canines

Dental Arcade: The rows of teeth in the upper and lower jaws.

- Chimpanzees have a narrow, U-shaped dental arcade

- Modern humans have a wider, parabolic dental arcade

- The dental arcade of Australopithecus afarensis has an intermediate V shape

Diastema: Functional gaps between teeth.

- In the chimpanzee's jaw, the gap between the canine and the neighboring incisor, which provides a space for the opposing canine when the animal's mouth is closed

- Modern humans have small canines and no diastema

9

Using Bones to Define Humans

Bipedalism

Fossil pelvic and leg bones, body proportions, and footprints all read "bipeds." The fossil bones are not identical to modern humans, but were likely functionally equivalent and a marked departure from those of quadrupedal chimpanzees.

Australopithecine fossils possess various components of the bipedal complex which can be compared to those of chimpanzees and humans:

- A diagnostic feature of bipedal locomotion is a shortened and broadened ilium; the australopithecine ilium is shorter than that of apes, and it is slightly curved; this shape suggests that the gluteal muscles were in a position to rotate and support the body during bipedal walking

- In modern humans, the head of the femur is robust, indicating increased stability at this joint for greater load bearing

- In humans, the femur angles inward from the hip to the knee joint, so that the lower limbs stand close to the body's midline. The line of gravity and weight are carried on the outside of the knee joint; in contrast, the chimpanzee femur articulates at the hip, then continues in a straight line downward to the knee joint

- The morphology of the australopithecine femur is distinct and suggests a slightly different function for the hip and knee joints. he femoral shaft is angled more than that of a chimpanzee and indicates that the knees and feet were well planted under the body

- In modern humans, the lower limbs bear all the body weight and perform all locomotor functions. Consequently, the hip, knee and ankle joint are all large with less mobility than their counterparts in chimpanzees. In australopithecines, the joints remain relatively small. In part, this might be due to smaller body size. It may also be due to a unique early hominid form of bipedal locomotion that differed somewhat from that of later hominids.

Thus human bodies were redesigned by natural selection for walking in an upright position for longer distances over uneven terrain.

Brain Size

Bipedal locomotion became established in the earliest stages of the hominid lineage, about 7 million years ago, whereas brain expansion came later. Early hominids had brains slightly larger than those of apes, but fossil hominids with significantly increased cranial capacities did not appear until about 2 million years ago.

Brain size remains near 450 cubic centimeters (cc) for robust australopithecines until almost 1.5 million years ago. At the same time, fossils assigned to Homo exceed 500 cc and reach almost 900 cc.

What might account for this later and rapid expansion of hominid brain size? One explanation is

10

called the "radiator theory": a new means for cooling this vital heat-generating organ, namely a new pattern of cerebral blood circulation, would be responsible for brain expansion in hominids. Gravitational forces on blood draining from the brain differ in quadrupedal animals versus bipedal animals: when humans stand bipedally, most blood drains into veins at the back of the neck, a network of small veins that form a complex system around the spinal column.

The two different drainage patterns might reflect two systems of cooling brains in early hominids. Active brains and bodies generate a lot of metabolic heat. The brain is a hot organ, but must maintain a fairly rigid temperature range to keep it functioning properly and to prevent permanent damage.

Savanna-dwelling hominids with this network of veins had a way to cool a bigger brain, allowing the "engine" to expand, contributing to hominid flexibility in moving into new habitats and in being active under a wide range of climatic conditions.

Free Hands

Unlike other primates, hominids no longer use their hands in locomotion or bearing weight or swinging through the trees. The chimpanzee's hand and foot are similar in size and length, reflecting the hand's use for bearing weight in knuckle walking. The human hand is shorter than the foot, with straighter phalanges. Fossil hand bones two million to three million years old reveal this shift in specialization of the hand from locomotion to manipulation.

Chimpanzee hands are a compromise. They must be relatively immobile in bearing weight during knuckle walking, but dexterous for using tools. Human hands are capable of power and precision grips but more importantly are uniquely suited for fine manipulation and coordination.

Tool Use

Fossil hand bones show greater potential for evidence of tool use. Although no stone tools are recognizable in an archaeological context until 2.5 million years ago, we can infer nevertheless their existence for the earliest stage of human evolution. The tradition of making and using tools almost certainly goes back much earlier to a period of utilizing unmodified stones and tools mainly of organic, perishable materials (wood or leaves) that would not be preserved in the fossil record.

How can we tell a hominid-made artifact from a stone generated by natural processes? First, the manufacturing process of hitting one stone with another to form a sharp cutting edge leaves a characteristic mark where the flake has been removed. Second, the raw material for the tools often comes from some distance away and indicates transport to the site by hominids.

Modification of rocks into predetermined shapes was a technological breakthrough. Possession of such tools opened up new possibilities in foraging: for example, the ability to crack open long bones and get at the marrow, to dig, and to sharpen or shape wooden implements.

Even before the fossil record of tools around 2.5 Myrs, australopithecine brains were larger than chimpanzee brains, suggesting increased motor skills and problem solving. All lines of evidence point to the importance of skilled making and using of tools in human evolution.

Summary

In this chapter, we learned the following:

1. Humans clearly depart from apes in several significant areas of anatomy, which stem from adaptation:

- bipedalism

- dentition (tooth size and shape)

- free hands

- brain size

2. For most of human evolution, cultural evolution played a fairly minor role. If we look back at the time of most australopithecines, it is obvious that culture had little or no influence on the lives of these creatures, who were constrained and directed by the same evolutionary pressures as the other organisms with which they shared their ecosystem. So, for most of the time during which hominids have existed, human evolution was no different from that of other organisms.

3. Nevertheless once our ancestors began to develop a dependence on culture for survival, then a new layer was added to human evolution. Sherwood Washburn suggested that the unique interplay of cultural change and biological change could account for why humans have become so different. According to him, as culture became more advantageous for the survival of our ancestors, natural selection favored the genes responsible for such behavior. These genes that improved our capacity for culture would have had an adaptive advantage. We can add that not only the genes but also anatomical changes made the transformations more advantageous. The ultimate result of the interplay between genes and culture was a significant acceleration of human evolution around 2.6 million to 2.5 million years ago.

EARLY HOMINID FOSSILS: REVIEW OF EVIDENCE

Overview of Human Evolutionary Origin

The fossil record provides little information about the evolution of the human lineage during the Late Miocene, from 10 million to 5 million years ago. Around 10 million years ago, several species of large-bodied hominoids that bore some resemblance to modern orangutans lived in Africa and Asia. About this time, the world began to cool; grassland and savanna habitats spread; and forests began to shrink in much of the tropics.

The creatures that occupied tropical forests declined in variety and abundance, while those that lived in the open grasslands thrived. We know that at least one ape species survived the environmental changes that occurred during the Late Miocene, because molecular genetics tells us that humans, gorillas, bonobos and chimpanzees are all descended from a common ancestor that lived sometime between 7 million and 5 million years ago. Unfortunately, the fossil record for the Late Miocene tells us little about the creature that linked the forest apes to modern hominids.

Beginning about 5 million years ago, hominids begin to appear in the fossil record. These early hominids were different from any of the Miocene apes in one important way: they walked upright (as we do). Otherwise, the earliest hominids were probably not much different from modern apes in their behavior or appearance.

Between 4 million and 2 million years ago, the hominid lineage diversified, creating a community of several hominid species that ranged through eastern and southern Africa. Among the members of this community, two distinct patterns of adaptation emerged:

- One group of creatures (Australopithecus, Ardipithecus, Paranthropus) evolved large molars that enhanced their ability to process coarse plant foods;

- The second group, constituted of members of our own genus Homo (as well as Australopithecus garhi) evolved larger brains, manufactured and used stone tools, and relied more on meat than the Australopithecines did.

Hominid Species

Species	Type	Specimen	Named by
Sahelanthropus	tchadensis "Toumai"	TM 266-01-060-1	Brunet et al. 2002
Orrorin	tugenensis	BAR 1000'00	Senut et al. 2001
Ardipithecus	ramidus	ARA-VP 6/1	White et al. 1994
Australopithecus	anamensis	KP 29281	M. Leakey et al. 1995
Australopithecus	afarensis	LH 4	Johanson et al. 1978
Australopithecus	bahrelghazali	KT 12/H1	Brunet et al. 1996
Kenyanthropus	platyops	KNM-WT 40000	M. Leakey et al. 2001
Australopithecus	garhi	BOU-VP-12/130	Asfaw et al. 1999
Australopithecus	africanus	Taung	Dart 1925
Australopithecus	aethiopicus	Omo 18	Arambourg & Coppens 1968
Paranthropus	robustus	TM 1517	Broom 1938
Paranthropus	boisei	OH 5	L. Leakey 1959
Homo	habilis	OH 7	L. Leakey et al. 1964

Sahelanthropus tchadensis ("Toumai")

- Named in July 2002 from fossils discovered in Chad.

- Oldest known hominid or near-hominid species (6-7 million years ago).

- Discovery of nearly complete cranium and number of fragmentary lower jaws and teeth:

- Skull has very small brain size (ca. 350 cc), considered as primitive apelike feature;

- Yet, other features are characteristic of later hominids: short and relatively flat face; canines are smaller and shorter; tooth enamel is slightly thicker (suggesting a diet with less fruit).

- This mixture of features, along with fact that it comes from around the time when hominids are thought to have diverged from chimpanzees, suggests it is close to the common ancestor of humans and chimpanzees.

- Foramen magnum is oval (not rounded as in chimps) suggesting upright walking position.

14

Orrorin tugenensis

- Named in July 2001; fossils discovered in western Kenya.

- Deposits dated to about 6 million years ago.

- Fossils include fragmentary arm and thigh bones, lower jaws, and teeth:

- Limb bones are about 1.5 times larger than those of Lucy, and suggest that it was about the size of a female chimpanzee.

- Its finders claimed that Orrorin was a human ancestor adapted to both bipedality and tree climbing, and that the australopithecines are an extinct offshoot.

Ardipithecus ramidus

- Recent discovery announced in Sept. 1994.

- Dated 4.4 million years ago.

- Most remains are skull fragments.

- Indirect evidence suggests that it was possibly bipedal, and that some individuals were about 122 cm (4'0") tall;

- Teeth are intermediate between those of earlier apes and Austalopithecus afarensis.

Australopithecus anamensis

- Named in August 1995 from fossils from Kanapoi and Allia Bay in Kenya.

- Dated between 4.2 and 3.9 million years ago.

- Fossils show mixture of primitive features in the skull, and advanced features in the body:

- Teeth and jaws are very similar to those of older fossil apes;

- Partial tibia is strong evidence of bipedality, and lower humerus (the upper arm bone) is extremely humanlike.

Australopithecus afarensis

- Existed between 3.9 and 3.0 million years ago.

- *A. afarensis* had an apelike face with a low forehead, a bony ridge over the eyes, a flat nose, and no chin. They had protruding jaws with large back teeth.

- Cranial capacity: 375 to 550 cc. Skull is similar to chimpanzee, except for more humanlike teeth. Canine teeth are much smaller than modern apes, but larger and more pointed than humans, and shape of the jaw is between rectangular shape of apes and parabolic shape of humans.

- Pelvis and leg bones far more closely resemble those of modern humans, and leave no doubt that they were bipedal.

- Bones show that they were physically very strong.

- Females were substantially smaller than males, a condition known as sexual dimorphism. Height varied between about 107 cm (3'6") and 152 cm (5'0").

- Finger and toe bones are curved and proportionally longer than in humans, but hands are similar to humans in most other details.

Kenyanthropus platyops ("flat-faced man of Kenya")

- Named in 2001 from partial skull found in Kenya.

- Dated to about 3.5 million years ago.

- Fossils show unusual mixture of features: size of skull is similar to *A. afarensis* and *A. africanus*, and has a large, flat face and small teeth.

Australopithecus garhi

- *A. garhi* existed around 2.5 Myrs.

- It has an apelike face in the lower part, with a protruding jaw resembling that of *A. afarensis*. The large size of the palate and teeth suggests that it is a male, with a small braincase of about 450 cc.

- It is like no other hominid species and is clearly not a robust form. In a few dental traits, such as the shape of the premolar and the size ratio of the canine teeth to the molars, *A. garhi* resembles specimens of early Homo. But its molars are huge, even larger than the *A. robustus* average.

- Among skeletal finds recovered, femur is relatively long, like that of modern humans. But forearm is long too, a condition found in apes and other australopithecines but not in humans.

Australopithecus africanus

- First identified in 1924 by Raymond Dart, an Australian anatomist living in South Africa.

- *A. africanus* existed between 3 and 2 million years ago.

- Similar to *A. afarensis*, and was also bipedal, but body size was slightly greater.

- Brain size may also have been slightly larger, ranging between 420 and 500 cc. This is a little larger than chimp brains (despite a similar body size).

- Back teeth were a little bigger than in *A. afarensis*. Although the teeth and jaws of *A. africanus* are much larger than those of humans, they are far more similar to human teeth than to those of apes. The shape of the jaw is now fully parabolic, like that of humans, and the size of the canine teeth is further reduced compared to *A. afarensis*.

NOTE: *Australopithecus afarensis* and *A. africanus* are known as gracile australopithecines, because of their relatively lighter build, especially in the skull and teeth. (Gracile means "slender", and in paleoanthropology is used as an antonym to "robust"). Despite the use of the word "gracile", these creatures were still more far more robust than modern humans.

Australopithecus aethiopicus

- *A. aethiopicus* existed between 2.6 and 2.3 million years ago.

- Species known mainly from one major specimen: the Black Skull (KNM-WT 17000) discovered at Lake Turkana.

- It may be ancestor of *P. robustus* and *P. boisei*, but it has a baffling mixture of primitive and advanced traits:

- Brain size is very small (410 cc) and parts of the skull (particularly the hind portions) are very primitive, most resembling *A. afarensis*;

- Other characteristics, like massiveness of face, jaws and largest sagittal crest in any known hominid, are more reminiscent of *P. boisei*.

Paranthropus boisei

- *P. boisei* existed between 2.2 and 1.3 million years ago.

- Similar to *P. robustus*, but face and cheek teeth were even more massive, some molars being up to 2 cm across. Brain size is very similar to P. robustus, about 530 cc.

- A few experts consider *P. boisei* and *P. robustus* to be variants of the same species.

Paranthropus robustus

- *P. robustus* had a body similar to that of *A. africanus*, but a larger and more robust skull and teeth.

- It existed between 2 and 1.5 million years ago.

- The massive face is flat, with large brow ridges and no forehead. It has relatively small front teeth, but massive grinding teeth in a large lower jaw. Most specimens have sagittal crests.

- Its diet would have been mostly coarse, tough food that needed a lot of chewing.

- The average brain size is about 530 cc. Bones excavated with *P. robustus* skeletons indicate that they may have been used as digging tools.

Australopithecus aethiopicus, Paranthropus robustus and *P. boisei* are known as robust australopithecines, because their skulls in particular are more heavily built.

Homo habilis

- *H. habilis* ("handy man") was so called because of evidence of tools found with its remains.

- *H. habilis* existed between 2.4 and 1.5 million years ago.

- It is very similar to australopithecines in many ways. The face is still primitive, but it projects less than in A. africanus. The back teeth are smaller, but still considerably larger than in modern humans.

- The average brain size, at 650 cc, is considerably larger than in australopithecines. Brain size varies between 500 and 800 cc, overlapping the australopithecines at the low end and *H. erectus* at the high end. The brain shape is also more humanlike.

- *H. habilis* is thought to have been about 127 cm (5'0") tall, and about 45 kg (100 lb) in weight, although females may have been smaller.

- Because of important morphological variation among the fossils, *H. habilis* has been a controversial species. Some scientists have not accepted it, believing that all *H. habilis* specimens should be assigned to either the australopithecines or Homo erectus. Many now believe that *H. habilis* combines specimens from at least two different *Homo* species: small-brained less-robust individuals (*H. habilis*) and large-brained, more robust ones (*H. rudolfensis*). Presently, not enough is known about these creatures to resolve this debate.

PHYLOGENY AND CHRONOLOGY

Between 8 million and 4 million years ago

Fossils of *Sahelanthropus tchadensis* (6-7 million years) and *Orrorin tugenensis* (6 million years), discovered in 2001 and 2000 respectively, are still a matter of debate.

The discoverers of *Orrorin tugenensis* claim the fossils represent the real ancestor of modern humans and that the other early hominids (e.g., Australopithecus and Paranthropus) are side branches. They base their claim on their assessment that this hominid was bipedal (2 million years earlier than previously thought) and exhibited expressions of certain traits that were more modern than those of other early hominids. Other authorities disagree with this analysis and some question whether this form is even a hominid. At this point, there is too little information to do more than mention these two new finds of hominids. As new data come in, however, a major part of our story could change.

Fossils of *Ardipithecus ramidus* (4.4 million years ago) were different enough from any found previously to warrant creating a new hominid genus. Although the evidence from the foramen magnum indicates that they were bipedal, conclusive evidence from legs, pelvis and feet remain somewhat enigmatic. There might be some consensus that *A. ramidus* represent a side branch of the hominid family.

Between 4 million and 2 million years ago

Australopithecus anamensis (4.2-3.8 million years ago) exhibit mixture of primitive (large canine teeth, parallel tooth rows) and derived (vertical root of canine, thicker tooth enamel) features, with evidence of bipedalism. There appears to be some consensus that this may represent the ancestor of all later hominids.

The next species is well established and its nature is generally agreed upon: *Australopithecus afarensis* (4-3 million years ago). There is no doubt that *A. afarensis* were bipeds. This form seems to still remain our best candidate for the species that gave rise to subsequent hominids.

At the same time lived a second species of hominid in Chad: *Australopithecus bahrelghazali* (3.5-3 million years ago). It suggests that early hominids were more widely spread on the African continent than previously thought. Yet full acceptance of this classification and the implications of the fossil await further study.

Another fossil species contemporaneous with A. afarensis existed in East Africa: *Kenyanthropus platyops* (3.5 million years ago). The fossils show a combination of features unlike that of any other forms: brain size, dentition, details of nasal region resemble genus Australopithecus; flat face, cheek area, brow ridges resemble later hominids. This set of traits led its discoverers to give it not only a new species name but a new genus name as well. Some authorities have suggested that this new form may be a better common ancestor for Homo than *A. afarensis*. More evidence and more examples with the same set of features, however, are needed to even establish that these fossils do represent a whole new taxonomy.

19

Little changed from A. afarensis to the next species: *A. africanus*: same body size and shape, and same brain size. There are a few differences, however: canine teeth are smaller, no gap in tooth row, tooth row more rounded (more human-like).

We may consider A. africanus as a continuation of *A. afarensis*, more widely distributed in southern and possibly eastern Africa and showing some evolutionary changes. It should be noted that this interpretation is not agreed upon by all investigators and remains hypothetical.

Fossils found at Bouri in Ethiopia led investigators to designate a new species: *A. garhi* (2.5 million years ago). Intriguing mixture of features: several features of teeth resemble early *Homo*; whereas molars are unusually larger, even larger than the southern African robust australopithecines.

The evolutionary relationship of A. garhi to other hominids is still a matter of debate. Its discoverers feel it is descended from A. afarensis and is a direct ancestor to Homo. Other disagree. Clearly, more evidence is needed to interpret these specimens more precisely, but they do show the extent of variation among hominids during this period.

Two distinctly different types of hominid appear between 2 and 3 million years ago: robust australopithecines (*Paranthropus*) and early *Homo* (*Homo habilis*).

The first type retains the chimpanzee-sized brains and small bodies of Australopithecus, but has evolved a notable robusticity in the areas of the skull involved with chewing: this is the group of robust australopithecines (*A. boisei, A. robustus, A. aethiopicus*).

- The Australopithecines diet seems to have consisted for the most part of plant foods, although *A. afarensis, A. africanus* and *A. garhi* may have consumed limited amounts of animal protein as well;

- Later Australopithecines (*A. boisei* and robustus) evolved into more specialized "grinding machines" as their jaws became markedly larger, while their brain size did not.

The second new hominid genus that appeared about 2.5 million years ago is the one to which modern humans belong, *Homo*.

- A Consideration of brain size relative to body size clearly indicates that Homo habilis had undergone enlargement of the brain far in excess of values predicted on the basis of body size alone. This means that there was a marked advance in information-processing capacity over that of Australopithecines;

- Although *H. habilis* had teeth that are large by modern standard, they are smaller in relation to the size of the skull than those of Australopithecines. Major brain-size increase and tooth-size reduction are important trends in the evolution of the genus Homo, but not of Australopithecines;

- From the standpoint of anatomy alone, it has long been recognized that either A. afarensis or *A. africanus* constitute a good ancestor for the genus Homo, and it now seems clear that the body of *Homo habilis* had changed little from that of either species. Precisely which of the two species gave rise to *H. habilis* is vigorously debated. Whether *H. habilis* is descended from *A. afarensis*, *A. africanus*, both of them, or neither of them, is still a matter of debate. It is also possible that none of the known australopithecines is our ancestor. The discoveries of *Sahelanthropus tchadensis*, *Orrorin tugenensis*, and *A. anamensis* are so recent that it is hard to say what effect they will have on current theories.

What might have caused the branching that founded the new forms of robust australopithecines (Paranthropus) and Homo? What caused the extinction, around the same time (between 2-3 million years ago) of genus Australopithecus? Finally, what might have caused the extinction of Paranthropus about 1 million years ago?

No certainty in answering these questions. But the environmental conditions at the time might hold some clues. Increased environmental variability, starting about 6 million years ago and continuing through time and resulting in a series of newly emerging and diverse habitats, may have initially promoted different adaptations among hominid populations, as seen in the branching that gave rise to the robust hominids and to Homo.

And if the degree of the environmental fluctuations continued to increase, this may have put such pressure on the hominid adaptive responses that those groups less able to cope eventually became extinct. Unable to survive well enough to perpetuate themselves in the face of decreasing resources (e.g., Paranthropus, who were specialized vegetarians) these now-extinct hominids were possibly out-competed for space and resources by the better adapted hominids, a phenomenon known as competitive exclusion.

In this case, only the adaptive response that included an increase in brain size, with its concomitant increase in ability to understand and manipulate the environment, proved successful in the long run.

HOMINOID, HOMINID, HUMAN

The traditional view has been to recognize three families of hominoid: the *Hylobatidae* (Asian lesser apes: gibbons and siamangs), the *Pongidae*, and the *Hominidae*.

- The Pongidae include the African great apes, including gorillas, chimpanzees, and the Asian orangutan;

- The Hominidae include living humans and typically fossil apes that possess a suite of characteristics such as bipedalism, reduced canine size, and increasing brain size (e.g., australopithecines).

The emergence of hominoids

Hominoids are Late Miocene (15-5 million years ago) primates that share a small number of postcranial features with living apes and humans:

- no tail;

- pelvis lacks bony expansion;

- elbow similar to that of modern apes;

- somewhat larger brains in relationship to body size than similarly sized monkeys.

When is a hominoid also a hominid?

When we say that *Sahelanthropus tchadensis* is the earliest hominid, we mean that it is the oldest fossil that is classified with humans in the family *Hominidae*. The rationale for including *Sahelanthropus tchadensis* in the *Hominidae* is based on similarities in shared derived characters that distinguish humans from other living primates.

There are three categories of traits that separate hominids from contemporary apes:

- bipedalism;

- much larger brain in relation to body size;

- dentition and musculature.

To be classified as a hominid, a Late Miocene primate (hominoid) must display at least some of these characteristics. *Sahelanthropus tchadensis* is bipedal, and shares many dental features with modern humans. However, the brain of *Sahelanthropus tchadensis* was no bigger than that of contemporary chimpanzees. As a consequence, this fossil is included in the same family (Hominidae) as modern humans, but not in the same genus. Imagine that.

Traits defining early *Homo*

Early Homo (e.g., *Homo habilis*) is distinctly different from any of the earliest hominids, including the australopithecines, and similar to us in the following ways:

- brain size is substantially bigger than that of any of the earliest hominids, including the australopithecines;

- teeth are smaller, enamel thinner, and the dental arcade is more parabolic than is found in the earliest hominids, including the australopithecines;

- skulls are more rounded; the face is smaller and protrudes less, and the jaw muscles are reduced compared with earliest hominids, including the australopithecines.

name=BEHAVIORAL_PATTERNS_OF_THE_EARLIEST_HOMINIDS

BEHAVIORAL PATTERNS OF THE EARLIEST HOMINIDS

One of the most important and intriguing questions in human evolution is about the diet of our earliest ancestors.

The presence of primitive stone tools in the fossil record tells us that 2.5 million years ago, early hominids (*A. garhi*) were using stone implements to cut the flesh off the bones of large animals that they had either hunted or whose carcasses they had scavenged.

Earlier than 2.5 million years ago, however, we know very little about the foods that the early hominids ate, and the role that meat played in their diet. Current situation due to lack of direct evidence.

Nevertheless, paleoanthropologists and archaeologists have tried to answer these questions indirectly using a number of techniques.

- Primatology (Studies on chimpanzee behavior)

- Anatomical Features (Tooth morphology and wear-patterns)

- Isotopic Studies

What does chimpanzee hunting behavior suggest about early hominid behavior?

Earliest ancestors and chimpanzees share a common ancestor (around 5-7 million years ago). Therefore, understanding chimpanzee hunting behavior and ecology may tell us a great deal about the behavior and ecology of those earliest hominids.

In the early 1960s, when Jane Goodall began her research on chimpanzees in Gombe National Park (Tanzania), it was thought that chimpanzees were strictly vegetarian. In fact, when Goodall first reported meat hunting by chimpanzees, many people were extremely skeptical.

Today, hunting by chimpanzees at Gombe and other locations in Africa has been well documented. We now know that each year chimpanzees may kill and eat more than 150 small and medium-sized animals, such as monkeys (red colobus monkey, their favorite prey), but also wild pigs and small antelopes.

Did early hominids hunt and eat small and medium-sized animals? It is quite possible that they did. We know that colobus-like monkeys inhabited the woodlands and riverside gallery forest in which early hominids lived 3-5 Myrs ago. There were also small animals and the young of larger animals to catch opportunistically on the ground. Many researchers now believe that the carcasses of dead animals were an important source of meat for early hominids once they had stone tools to use (after 2.5 million years ago) for removing the flesh from the carcass. Wild chimpanzees show little interest in dead animals as a food source, so scavenging may have

24

evolved as an important mode of getting food when hominids began to make and use tools for getting at meat. Before this time, it seems likely that earlier hominids were hunting small mammals as chimpanzees do today and that the role that hunting played in the early hominids' social lives was probably as complex and political as it is in the social lives of chimpanzees.

When we ask when meat became an important part of the human diet, we therefore must look well before the evolutionary split between apes and humans in our own family tree.

What do tooth wear patterns suggest about early hominid behavior?

Bones and teeth in the living person are very plastic and respond to mechanical stimuli over the course of an individual's lifetime. We know, for example, that food consistency (hard vs. soft) has a strong impact on the masticatory (chewing) system (muscles and teeth). Bones and teeth in the living person are therefore tissues that are remarkably sensitive to the environment. As such, human remains from archaeological sites offer us a retrospective biological picture of the past that is rarely available from other lines of evidence. Also, new technological advances developed in the past ten years or so now make it possible to reconstruct and interpret in amazing detail the physical activities and adaptations of hominids in diverse environmental settings.

Some types of foods are more difficult to process than others, and primates tend to specialize in different kinds of diets. Most living primates show three basic dietary adaptations:

- insectivores (insect eaters);

- frugivores (fruit eaters);

- folivores (leaf eaters)

Many primates, such as humans, show a combination of these patterns and are called omnivores, which in a few primates includes eating meat.

The ingestion both of leaves and of insects requires that the leaves and the insect skeletons be broken up and chopped into small pieces. The molars of folivores and insectivores are characterized by the development of shearing crests on the molars that function to cut food into small pieces. Insectivores' molars are further characterized by high, pointed cusps that are capable of puncturing the outside skeleton of insects. Frugivores, on the other hand, have molar teeth with low, rounded cusps; their molars have few crests and are characterized by broad, flat basins for crushing the food.

In the 1950s, John Robinson developed what came to be known as the dietary hypothesis. According to this theory there were fundamentally two kinds of hominids in the Plio-Pleistocene. On was the "robust" australopithecine (called Paranthropus) that was specialized for herbivory, and the other was the "gracile" australopithecine that was an omnivore/carnivore. By this theory the former became extinct while the latter evolved into *Homo*.

Like most generalizations about human evolution, Robinson's dietary hypothesis was controversial, but it stood as a useful model for decades.

Detailed analyses of the tooth surface under microscope appeared to confirm that the diet of *A. robustus* consisted primarily of plants, particularly small and hard objects like seeds, nuts and tubers. The relative sizes and shapes of the teeth of both *A. afarensis* and *A. africanus* indicated as well a mostly mixed vegetable diet of fruits and leaves. By contrast, early *Homo* was more omnivorous.

But as new fossil hominid species were discovered in East Africa and new analyses were done on the old fossils, the usefulness of the model diminished.

For instance, there is a new understanding that the two South African species (*A. africanus* and *A. robustus*) are very similar when compared to other early hominid species. They share a suite of traits that are absent in earlier species of Australopithecus, including expanded cheek teeth and faces reemodeled to withstand forces generated from heavy chewing.

What do isotopic studies suggest about early hominid behavior?

Omnivory can be suggested by studies of the stable carbon isotopes and strontium(Sr)-calcium(Ca) ratios in early hominid teeth and bones.

For instance, a recent study of carbon isotope (13C) in the tooth enamel of a sample of *A. africanus* indicated that members of this species ate either tropical grasses or the flesh of animals that ate tropical grasses or both. But because the dentition analyzed by these researchers lacked the tooth wear patterns indicative of grass-eating, the carbon may have come from grass-eating animals. This is therefore a possible evidence that the australopithecines either hunted small animals or scavenged the carcasses of larger ones.

There is new evidence also that *A. robustus* might not be a strict vegetarian. Isotopic studies reveal chemical signals associated with animals whose diet is omnivorous and not specialized herbivory. The results from 13C analysis indicate that *A. robustus* either ate grass and grass seeds or ate animals that ate grasses. Since the Sr/Ca ratios suggest that *A. robustus* did not eat grasses, these data indicate that *A. robustus* was at least partially carnivorous.

Summary

Much of the evidence for the earliest hominids (*Sahelanthropus tchadensis*, *Orrorin tugenensis*, *Ardipithecus ramidus*) is not yet available.

Australopithecus anamensis shows the first indications of thicker molar enamel in a hominid. This suggests that *A. anamensis* might have been the first hominid to be able to effectively withstand the functional demands of hard and perhaps abrasive objects in its diet, whether or not such items were frequently eaten or were only an important occasional food source.

Australopithecus afarensis was similar to *A. anamensis* in relative tooth sizes and probable enamel thickness, yet it did show a large increase in mandibular robusticity. Hard and perhaps abrasive foods may have become then even more important components of the diet of *A. afarensis*.

Australopithecus africanus shows yet another increase in postcanine tooth size, which in itself would suggest an increase in the sizes and abrasiveness of foods. However, its molar microwear does not show the degree of pitting one might expect from a classic hard-object feeder. Thus, even *A. africanus* has evidently not begun to specialize in hard objects, but rather has emphasized dietary breadth (omnivore), as evidenced by isotopic studies.

Subsequent "robust" australopithecines do show hard-object microwear and craniodental specializations, suggesting a substantial departude in feeding adaptive strategies early in the Pleistocene. Yet, recent chemical and anatomical studies on *A. robustus* suggest that this species may have consumed some animal protein. In fact, they might have specialized on tough plant material during the dry season but had a more diverse diet during the rest of the year.

THE OLDOWAN PERIOD

The Olduvai Gorge

2 million years ago, Olduvai Gorge (Tanzania) was a lake. Its shores were inhabited not only by numerous wild animals but also by groups of hominids, including Paranthropus boisei and Homo habilis, as well as the later Homo erectus.

The gorge, therefore, is a great source of Palaeolithic remains as well as a key site providing evidence of human evolutionary development. This is one of the main reasons that drew Louis and Mary Leakey back year after year at Olduvai Gorge.

Certain details of the lives of the creatures who lived at Olduvai have been reconstructed from the hundreds of thousands of bits of material that they left behind: various stones and bones. No one of these things, alone, would mean much, but when all are analyzed and fitted together, patterns begin to emerge.

Among the finds are assemblages of stone tools dated to between 2.2 Myrs and 620,000 years ago. These were found little disturbed from when they were left, together with the bones of now-extinct animals that provided food.

Mary Leakey found that there were two stoneworking traditions at Olduvai. One, the Acheulean industry, appears first in Bed II and lasts until Bed IV. The other, the Oldowan, is older and more primitive, and occurs throughout Bed I, as well as at other African sites in Ethiopia, Kenya and Tanzania.

Subsistence patterns

Meat-eating

Until about 2.5 million years ago, early hominids lived on foods that could be picked or gathered: plants, fruits, invertebrate animals such as ants and termites, and even occasional pieces of meat (perhaps hunted in the same manner as chimpanzees do today).

After 2.5 million years ago, meat seems to become more important in early hominids' diet. Evolving hominids' new interest in meat is of major importance in paleoanthropology.

Out on the savanna, it is hard for a primate with a digestive system like that of humans to satisfy its amino-acid requirements from available plant resources. Moreover, failure to do so has serious consequences: growth depression, malnutrition, and ultimately death. The most readily accessible plant resources would have been the proteins accessible in leaves and legumes, but these are hard for primates like us to digest unless they are cooked. In contrast, animal foods (ants, termites, eggs) not only are easily digestible, but they provide high-quantity proteins that contain all the essential amino acids. All things considered, we should not be surprised if our own ancestors solved their "protein problem" in somewhat the same way that chimps on the savanna do today.

Increased meat consumption on the part of early hominids did more than merely ensure an adequate intake of essential amino acids. Animals that live on plant foods must eat large quantities of vegetation, and obtaining such foods consumes much of their time. Meat eaters, by contrast, have no need to eat so much or so often. Consequently, meat-eating hominids may have had more leisure time available to explore and manipulate their environment, and to lie around and play. Such activities probably were a stimulus to hominid brain development.

The importance of meat eating for early hominid brain development is suggested by the size of their brains:

- cranial capacity of largely plant-eating Australopithecus ranged from 310 to 530 cc;

- cranial capacity of primitive known meat eater, Homo habilis: 580 to 752 cc;

- *Homo erectus* possessed a cranial capacity of 775 to 1,225 cc.

Hunters or scavengers?

The archaeological evidence indicates that Oldowan hominids ate meat. They processed the carcasses of large animals, and we assume that they ate the meat they cut from the bones. Meat-eating animals can acquire meat in several different ways:

- stealing kills made by other animals;

- by opportunistically exploiting the carcasses of animals that die naturally;

- by hunting or capturing prey themselves.

There has been considerable dispute among anthropologists about how early hominids acquired meat. Some have argued that hunting, division of labor, use of home bases and food sharing emerged very early in hominid history. Others think the Oldowan hominids would have been unable to capture large mammals because they were too small and too poorly armed.

Recent zooarchaeological evidence suggests that early hominids (after 2.5 million years ago) may have acquired meat mainly by scavenging, and maybe occasionally by hunting.

If hominids obtained most of their meat from scavenging, we would expect to find cut marks mainly on bones left at kill sites by predators (lions, hyenas). If hominids obtained most of their meat from their own kills, we would expect to find cut marks mainly on large bones, like limb bones. However, at Olduvai Gorge, cut marks appear on both kinds of bones: those usually left by scavengers and those normally monopolized by hunters. The evidence from tool marks on bones indicates that humans sometimes acquired meaty bones before, and sometimes after, other predators had gnawed on them.

Settlement patterns

During decades of work at Olduvai Gorge, Mary and Louis Leakey and their team laid bare numerous ancient hominid sites. Sometimes the sites were simply spots where the bones of one or more hominid species were discovered. Often, however, hominid remains were found in association with concentrations of animal bones, stone tools, and debris.

At one spot, in Bed I, the bones of an elephant lay in close association with more than 200 stone tools. Apparently, the animal was butchered here; there are no indications of any other activity.

At another spot (DK-I Site), on an occupation surface 1.8 million years old, basalt stones were found grouped in small heaps forming a circle. The interior of the circle was practically empty, while numerous tools and food debris littered the ground outside, right up to the edge of the circle.

Earliest stone industry

Principles

Use of specially made stone tools appears to have arisen as result of need for implements to butcher and prepare meat, because hominid teeth were inadequate for the task. Transformation of lump of stone into a "chopper", "knife" or "scraper" is a far cry from what a chimpanzee does when it transforms a stick into a termite probe. The stone tool is quite unlike the lump of stone. Thus, the toolmaker must have in mind an abstract idea of the tool to be made, as well as a specific set of steps that will accomplish the transformation from raw material to finished product. Furthermore, only certain kinds of stone have the flaking properties that will allow the transformation to take place, and the toolmaker must know about these.

Therefore, two main components to remember:

- Raw material properties
- Flaking properties

Evidence

The oldest Lower Palaeolithic tools (2.0-1.5 million years ago) found at Olduvai Gorge (*Homo habilis*) are in the Oldowan tool tradition. Nevertheless, older materials (2.6-2.5 million year ago) have recently been recorded from sites located in Ethiopia (Hadar, Omo, Gona, Bouri - *Australopithecus garhi*) and Kenya (Lokalalei).

Because of a lack of remarkable differences in the techniques and styles of artifact manufacture for over 1 million years (2.6-1.5 million years ago), a technological stasis was suggested for the Oldowan Industry.

The makers of the earliest stone artifacts travelled some distances to acquire their raw materials, implying greater mobility, long-term planning and foresight not recognized earlier.

Oldowan stone tools consist of all-purpose generalized chopping tools and flakes. Although these artifacts are very crude, it is clear that they have been deliberately modified. The technique of manufacture used was the percussion.

The main intent of Oldowan tool makers was the production of cores and flakes with sharp-edges. These simple but effective Oldowan choppers and flakes made possible the addition of meat to the diet on a regular basis, because people could now butcher meat, skin any animal, and split bones for marrow.

Overall, the hominids responsible for making these stone tools understood the flaking properties of the raw materials available; they selected appropriate cobbles for making artefacts; and they were as competent as later hominids in their knapping abilities.

Finally, the manufacture of stone tools must have played a major role in the evolution of the human brain, first by putting a premium on manual dexterity and fine manipulation over mere power in the use of the hands. This in turn put a premium in the use of the hands.

Early hominid behavior

During the 1970s and 1980s many workers, including Mary Leakey and Glynn Isaac, used an analogy from modern hunter-gatherer cultures to interpret early hominid behavior of the Oldowan period (e.g., the Bed I sites at Olduvai Gorge). They concluded that many of the sites were probably camps, often called "home bases", where group members gathered at the end of the day to prepare and share food, to socialize, to make tools, and to sleep.

The circular concentration of stones at the DK-I site was interpreted as the remains of a shelter or windbreak similar to those still made by some African foraging cultures. Other concentrations of bones and stones were thought to be the remains of living sites originally ringed by thorn hedges for defense against predators. Later, other humanlike elements were added to the mix, and early Homo was described as showing a sexual division of labor [females gathering plant foods and males hunting for meat] and some of the Olduvai occupation levels were interpreted as butchering sites.

Views on the lifestyle of early *Homo* began to change in the late 1980s, as many scholars became convinced that these hominids had been overly humanized.

Researchers began to show that early *Homo* shared the Olduvai sites with a variety of large carnivores, thus weakening the idea that these were the safe, social home bases originally envisioned.

Studies of bone accumulations suggested that *H. habilis* was mainly a scavenger and not a full-fledged hunter. The bed I sites were interpreted as no more than "scavenging stations" where early *Homo* brought portions of large animal carcasses for consumption.

Another recent suggestion is that the Olduvai Bed I sites mainly represent places where rocks were cached for the handy processing of animal foods obtained nearby. Oldowan toolmakers brought stones from sources several kilometers away and cached them at a number of locations within the group's territory. Stone tools could have been made at the cache sites for use elsewhere, but more frequently portions of carcasses were transported to the toolmaking site for processing.

31

Summary

Current interpretations of the subsistence, settlement, and tool-use patterns of early hominids of the Oldowan period are more conservative than they have been in the past. Based upon these revised interpretations, the Oldowan toolmakers have recently been dehumanized.

Although much more advanced than advanced apes, they still were probably quite different from modern people with regard to their living arrangements, methods and sexual division of food procurement and the sharing of food.

The label human has to await the appearance of the next representative of the hominid family: *Homo erectus*.

THE ACHEULIAN PERIOD

In 1866, German biologist Ernst Haeckel had proposed the generic name "*Pithecanthropus*" for a hypothetical missing link between apes and humans.

In late 19th century, Dutch anatomist Eugene Dubois was in Indonesia, precisely on the Island of Java, in search for human fossils. Between 1887 and 1892, he encountered various fragments of skulls and long bones which convinced him he had discovered an erect, apelike transitional form between apes and humans. In 1894, he decided to call his fossil species *Pithecanthropus erectus*. Dubois found no additional human fossils and he returned to the Netherlands in 1895.

Others explored the same deposits on the Island of Java, but new human remains appeared only between 1931 and 1933.

Dubois's claim for a primitive human species was further reinforced by nearly simultaneous discoveries from near Beijing, China (at the site of Zhoukoudian). Between 1921 and 1937, various scholars undertook fieldwork in one collapsed cave (Locality 1) recovered many fragments of mandibles and skulls. One of them, Davidson Black, a Canadian anatomist, created a new genus and species for these fossils: *Sinanthropus pekinensis* ("Peking Chinese man").

In 1939, after comparison of the fossils in China and Java, some scholars concluded that they were extremely similar. They even proposed that Pithecanthropus and Sinanthropus were only subspecies of a single species, *Homo erectus*, though they continued to used the original generic names as labels.

From 1950 to 1964, various influential authorities in paleoanthropology agreed that Pithecanthropus and Sinanthropus were too similar to be placed in two different genera, and, by the late 1960s, the concept of Homo erectus was widely accepted.

To the East Asian inventory of *H. erectus*, many authorities would add European and especially African specimens that resembled the Asian fossil forms. In 1976, a team led by Richard Leakey discovered around Lake Turkana (Kenya) an amazingly well preserved and complete skeleton of a *H. erectus* boy, called the Turkana Boy (WT-15000).

In 1980s and 1990s:

- new discoveries in Asia (Longgupo, Dmanisi, etc.); in Europe (Atapuerca, Orce, Ceprano);

- precision in chronology and evolution of *H. erectus*;

- understanding and definition of variability of this species and relationship with other contemporary species.

Site distribution

Africa

Unlike Australopithecines and even *Homo habilis*, *Homo ergaster/erectus* was distributed throughout Africa:

- about 1.5 million years ago, shortly after the emergence of *H. ergaster*, people more intensively occupied the Eastern Rift Valley;

- by 1 million years ago, they had extended their range to the far northern and southern margins of Africa.

Traditionally, Homo erectus has been credited as being the prehistoric pioneer, a species that left Africa about 1 million years ago and began to disperse throughout Eurasia. But several important discoveries in the 1990s have reopened the question of when our ancestors first journeyed from Africa to other parts of the globe. Recent evidence now indicates that emigrant erectus made a much earlier departure from Africa.

Israel

Ubeidiyeh

- Deposits accumulated between 1.4-1.0 million years ago;

- Stone tools of both an early chopper-core (or Developed Oldowan) industry and crude Acheulean-like handaxes. The artifacts closely resemble contemporaneous pieces from Upper Bed II at Olduvai Gorge;

- Rare hominid remains attributed to Homo erectus;

- Ubeidiya might reflect a slight ecological enlargement of Africa more than a true human dispersal.

Gesher Benot Yaaqov

- 800,000 years ago;

- No hominid remains;

- Stone tools are of Acheulean tradition and strongly resemble East African industries.

Republic of Georgia

In 1991, archaeologists excavating a grain-storage pit in the medieval town of Dmanisi uncovered the lower jaw of an adult erectus, along with animal bones and Oldowan stone tools.

Different dating techniques (paleomagnetism, potassium-argon) gave a date of 1.8 million years ago, that clearly antedate that of Ubeidiya. Also the evidence from Dmanisi suggests now a true migration from Africa.

China

Longgupo Cave

- Dated to 1.8 million years ago

- Fragments of a lower jaw belonging either to Homo erectus or an unspecified early *Homo*.

- Fossils recovered with Oldowan tools.

Zhoukoudian

- Dated between 500,000 and 250,000 years ago.

- Remarkable site for providing large numbers of fossils, tools and other artifacts.

- Fossils of *Homo erectus* discovered in 1920s and 1930s.

Java

In 1994, report of new dates from sites of Modjokerto and Sangiran where *H. erectus* had been found in 1891.

Geological age for these hominid remains had been estimated at about 1 million years old. Recent redating of these materials gave dates of 1.8 milllion years ago for the Modjokerto site and 1.6 milllion years ago for the Sangiran site.

These dates remained striking due to the absence of any other firm evidence for early humans in East Asia prior to 1 Myrs ago. Yet the individuals from Modjokerto and Sangiran would have certainly traveled through this part of Asia to reach Java.

Europe

Did Homo ergaster/erectus only head east into Asia, altogether bypassing Europe?

Many paleoanthropologists believed until recently that no early humans entered Europe until 500,000 years ago. But the discovery of new fossils from Spain (Atapuerca, Orce) and Italy (Ceprano) secured a more ancient arrival for early humans in Europe.

At Atapuerca, hundreds of flaked stones and roughly eighty human bone fragments were collected from sediments that antedate 780,000 years ago, and an age of about 800,000 years ago is the current best estimate. The artifacts comprise crudely flaked pebbles and simple flakes. The hominid fossils - teeth, jaws, skull fragments - come from several individuals of a new species named Homo antecessor. These craniofacial fragments are striking for derived features that differentiate them from Homo ergaster/erectus, but do not ally them specially with either *H.*

neanderthalensis or *H. sapiens*.

HOMINIDS OF THE ACHEULIAN PERIOD

The hominids

AFRICAN *HOMO ERECTUS*: *HOMO ERGASTER*

H. ergaster existed between 1.8 million and 1.3 million years ago.

Like *H. habilis*, the face shows:

- protruding jaws with large molars;

- no chin;

- thick brow ridges;

- long low skull, with a brain size varying between 750 and 1225 cc.

Early *H. ergaster* specimens average about 900 cc, while late ones have an average of about 1100 cc. The skeleton is more robust than those of modern humans, implying greater strength.

Body proportions vary:

Ex. Turkana Boy is tall and slender, like modern humans from the same area, while the few limb bones found of Peking Man indicate a shorter, sturdier build.

Study of the Turkana Boy skeleton indicates that H. ergaster may have been more efficient at walking than modern humans, whose skeletons have had to adapt to allow for the birth of larger brained infants.

Homo habilis and all the australopithecines are found only in Africa, but H. erectus/ergaster was wide-ranging, and has been found in Africa, Asia, and Europe.

ASIAN *HOMO ERECTUS*

Specimens of *H. erectus* from Eastern Asia differ morphologically from African specimens:

- features are more exaggerated;

- skull is thicker, brow ridges are more pronounced, sides of skull slope more steeply, the sagittal crest is more exaggerated;

- Asian forms do not show the increase in cranial capacity.

As a consequence of these features, they are less like humans than the African forms of *H. erectus*.

Paleoanthropologists who study extinct populations are forced to decide whether there was one species or two based on morphological traits alone. They must ask whether eastern and western forms are as different from each other as typical species.

If systematics finally agree that eastern and western populations of *H. erectus* are distinct species, then the eastern Asian form will keep the name *H. erectus*. The western forms have been given a new name: *Homo ergaster* (means "work man") and was first applied to a very old specimen from East Turkana in East Africa.

HOMO GEORGICUS

Specimens recovered recently exhibit characteristic *H. erectus* features: sagittal crest, marked constriction of the skull behind the eyes. But they are also extremely different in several ways, resembling *H. habilis*:

- small brain size (600 cc);

- prominent browridge;

- projection of the face;

- rounded contour of the rear of skull;

- huge canine teeth.

Some researchers propose that these fossils might represent a new species of Homo: *H. georgicus*.

HOMO ANTECESSOR

Named in 1997 from fossils (juvenile specimen) found in Atapuerca (Spain). Dated to at least 780,000 years ago, it makes these fossils the oldest confirmed European hominids.

Mid-facial area of antecessor seems very modern, but other parts of skull (e.g., teeth, forehead and browridges) are much more primitive. Fossils assigned to new species on grounds that they exhibit unknown combination of traits: they are less derived in the Neanderthal direction than later mid-Quaternary European specimens assigned to Homo heidelbergensis.

HOMO HEIDELBERGENSIS

Archaic forms of Homo sapiens first appeared in Europe about 500,000 years ago (until about 200,000 years ago) and are called Homo heidelbergensis.

Found in various places in Europe, Africa and maybe Asia.

This species covers a diverse group of skulls which have features of both Homo erectus and modern humans.

Fossil features:

- brain size is larger than erectus and smaller than most modern humans: averaging about 1200 cc;

- skull is more rounded than in erectus;

- still large brow ridges and receding foreheads;

- skeleton and teeth are usually less robust than erectus, but more robust than modern humans;

- mandible is human-like, but massive and chinless; shows expansion of molar cavities and very long cheek tooth row, which implies a long, forwardly projecting face.

Fossils could represent a population near the common ancestry of Neanderthals and modern humans.

Footprints of H. heidelbergensis (earliest human footprints) have been found in Italy in 2003.

Phylogenic Relationships

For almost three decades, paleoanthropologists have often divided the genus Homo among three successive species:

- *Homo habilis*, now dated between roughly 2.5 Myrs and 1.7 Myrs ago;

- *Homo erectus*, now placed between roughly 1.7 Myrs and 500,000 years ago;

- *Homo sapiens*, after 500,000 years ago.

In this view, each species was distinguished from its predecessor primarily by larger brain size and by details of cranio-facial morphology:

Ex. Change in braincase shape from more rounded in *H. habilis* to more angular in H. erectus to more rounded again in *H. sapiens*.

The accumulating evidence of fossils has increasingly undermined a scenario based on three successive species or evolutionary stages. It now strongly favors a scheme that more explicitly recognizes the importance of branching in the evolution of Homo.

This new scheme continues to accept *H. habilis* as the ancestor for all later Homo. Its descendants at 1.8-1.7 million mears ago may still be called H. erectus, but H. ergaster is now more widely accepted. By 600,000-500,000 years ago, *H. ergaster* had produced several lines leading to H. neanderthalensis in Europe and *H. sapiens* in Africa. About 600,000 years ago, both of these species shared a common ancestor to which the name H. heidelbergensis could be applied.

"Out-of-Africa 1" model

Homo erectus in Asia would be as old as Homo ergaster in Africa. Do the new dates from Dmanisi and Java falsify the hypothesis of an African origin for *Homo erectus*? Not necessarily.

If the species evolved just slightly earlier than the oldest African fossils (2.0-1.9 million years ago) and then immediately began its geographic spread, it could have reached Europe and Asia fairly quickly.

But the "Out-of-Africa 1" migration is more complex. Conventional paleoanthropological wisdom holds that the first human to leave Africa were tall, large-brained hominids (*Homo ergaster/erectus*). New fossils discovered in Georgia (Dmanisi) are forcing scholars to rethink that scenario completely. These Georgian hominids are far smaller and more primitive in both anatomy and technology than expected, leaving experts wondering not only why early humans first ventured out of Africa, but also how.

Summary

Homo ergaster was the first hominid species whose anatomy fully justify the label human:

- Unlike australopithecines and *Homo habilis*, in which body form and proportions retained apelike features suggesting a continued reliance on trees for food or refuge, *H. ergaster* achieved essentially modern forms and proportions;

- Members also differed from australopithecines and *H. habilis* in their increased, essentially modern stature and in their reduced degree of sexual dimorphism.

ACHEULEAN TECHNOLOGY AND SUBSISTENCE

Homo ergaster/erectus, the author of the Acheulean industry, enjoyed impressive longevity as a species and great geographic spread. We will review several cultural innovations and behavioral changes that might have contributed to the success of *H. ergaster/erectus*:

- stone-knapping advances that resulted in Acheulean bifacial tools;

- the beginnings of shelter construction;

- the control and use of fire;

- increased dependence on hunting.

The Acheulean industrial complex

(1.7 million - 200,000 years ago)

Stone tools

By the time *Homo ergaster/erectus* appeared, Oldowan choppers and flake tools had been in use for 800,000 years. For another 100,000 to 400,000 years, Oldowan tools continued to be the top-of-the-line implements for early *Homo ergaster/erectus*. Between 1.7 and 1.4 million years ago, Africa witnessed a significant advance in stone tool technology: the development of the Acheulean industry.

The Acheulean tool kit included:

- picks;

- cleavers;

- an assortment of Oldowan-type choppers and flakes, suggesting that the more primitive implements continued to serve important functions;

- mainly characterized by bifacially flaked tools, called bifaces.

A biface reveals a cutting edge that has been flaked carefully on both sides to make it straighter and sharper than the primitive Oldowan chopper. The purpose of the two-sided, or bifacial, method was to change the shape of the core from essentially round to flattish, for only with a flat stone can one get a decent cutting edge.

One technological improvement that permitted the more controlled working required to shape an Acheulean handax was the gradual implementation, during the Acheulean period, of different kinds of hammers. In earlier times, the toolmaker knocked flakes from the core with another piece of stone. The hard shock of rock on rock tended to leave deep, irregular scars and wavy cutting edges.

But a wood or bone hammer, being softer, gave its user much greater control over flaking. Such implements left shallower, cleaner scars and produced sharper and straighter cutting edges.

With the Acheulean Industry, the use of stone (hard hammer) was pretty much restricted to the preliminary rough shaping of a handax, and all the fine work around the edges was done with wood and bone.

Acheulean handaxes and cleavers are generally interpreted as being implements for processing animal carcasses. Even though cleavers could have been used to chop and shape wood, their wear patterns are more suggestive of use on soft material, such as hides and meat. Acheulean tools represent an adaptation for habitual and systematic butchery, and especially the dismembering of large animal carcasses, as *Homo ergaster/erectus* experienced a strong dietary shift toward more meat consumption.

Acheulean tools originated in Africa between 1.7 and 1.4 million years ago. They were then produced continuously throughout *Homo ergaster/erectus'* long African residency and beyond, finally disappearing about 200,000 years ago.

Generally, Acheulean tools from sites clearly older than 400,000 to 500,000 years ago are attributed to Homo ergaster/erectus, even in the absence of confirming fossils. At several important Late Acheulean sites, however, the toolmakers' species identity remains ambiguous because the sites lack hominid fossils and they date to a period when *Homo erectus* and archaic *Homo sapiens* (e.g., *Homo heidelbergensis*) overlapped in time.

Other raw matierials

Stone artifacts dominate the Paleolithic record because of their durability, but early people surely used other raw materials, including bone and more perishable substances like wood, reeds, and skin.

A few sites, mainly European, have produced wooden artifacts, which date usually between roughly 600,000 and 300,000 years ago:

Ex. At the site of SchÃ¶ningen, Germany, several wooden throwing spears, over 2 m long. They arguably present the oldest, most compelling case for early human hunting.

Diffusion of Technology

Wide variability in stone tools present with *H. erectus*. In Eastern Asia, H. erectus specimens are associated not with Acheulean tools, but instead with Oldowan tools, which were retained until 200,000 to 300,000 years ago.

This pattern was first pointed out by Hallam Movius in 1948. The line dividing the Old World into Acheulean and non-Acheulean regions became known as the Movius line. Handax cultures flourished to the west and south of the line, but in the east, only choppers and flake tools were found.

Why were there no Acheulean handax cultures in the Eastern provinces of Asia?

- history of research;

- other explanations:

- quality of raw materials (fine-grained rocks rare)

- different functional requirements (related to

- environment and food procurement)

- "bamboo culture": bamboo tools used in place of

- stone implements to perform tasks;

- Early dates in Java and Dmanisi explain situation

  Acheulean developed in Africa from the preceding Oldowan Tradition only after 1.8 Myrs ago, but if people moved into eastern Asia at 1.8 million years ago or before, they would have arrived without Acheulean tools.

  In sum, while the Acheulean tradition, with its handaxes and cleavers, was an important lithic advance by Homo ergaster over older technologies, it constituted only one of several adaptive patterns used by the species. Clever and behaviorally flexible, H. ergaster was capable of adjusting its material culture to local resources and functional requirements.

Subsistence patterns and diet

Early discoveries of Homo ergaster/erectus fossils in association with stone tools and animal bones lent themselves to the interpretation of hunting and gathering way of life. Nevertheless this interpretation is not accepted by all scholars and various models have been offered to make sense of the evidence.

First Scenario: Scavenging

Recently, several of the original studies describing *Homo ergaster/erectus* as a hunter-gatherer have come under intense criticism. Re-examination of the material at some of the sites convinced some scholars (L. Binford) that faunal assemblages were primarily the result of animal activity rather than hunting and gathering.

Animal bones showed cut marks from stone tools that overlay gnaw marks by carnivores, suggesting that *Homo ergaster/erectus* was not above scavenging parts of a carnivore kill.

According to these scholars, at most sites, the evidence for scavenging by hominids is much more convincing than is that for actual hunting.

Which scenario to choose?

The key point here is not that *Homo ergaster/erectus* were the first hominid hunters, but that they depended on meat for a much larger portion of their diet than had any previous hominid species.

Occasional hunting is seen among nonhuman primates and cannot be denied to australopithecines (see *A. garhi*). But apparently for *Homo ergaster/erectus* hunting took an unprecedented importance, and in doing so it must have played a major role in shaping both material culture and society.

Shelter and fire

For years, scientists have searched for evidence that *Homo ergaster/erectus* had gained additional controlled over its environment through the construction of shelters, and the control and use of fire. The evidence is sparse and difficult to interpret.

Shelter

Seemingly patterned arrangements or concentrations of large rocks at sites in Europe and Africa may mark the foundations of huts or windbreaks, but in each case the responsible agent could equally well be stream flow, or any other natural process.

Therefore there appears to be no convincing evidence that *Homo ergaster/erectus* regularly constructed huts, windbreaks, or any other sort of shelter during the bulk of its long period of existence. Shelter construction apparently developed late in the species' life span, if at all, and therefore cannot be used as an explanation of *H. ergaster's* capacity for geographic expansion.

Fire

Proving the evidence of fire by *Homo ergaster/erectus* is almost equally problematic. Some researchers have suggested that the oldest evidence for fire use comes from some Kenyan sites dated about 1.4 to 1.6 million years ago. Other scholars are not sure. The problem is that the baked earth found at these sites could have been produced as easily by natural fires as by fires started - or at least controlled - by *H. ergaster/erectus*.

Better evidence of fire use comes from sites that date near the end of *Homo erectus'* existence as a species. Unfortunately, the identity of the responsible hominids (either Homo erectus or archaic Homo sapiens) is unclear.

The evidence at present suggests that fire was not a key to either the geographic spread or the longevity of these early humans.

Out-of-Africa 1: Behavioral aspects

Researchers proposed originally that it was not until the advent of handaxes and other symmetrically shaped, standardized stone tools that *H. erectus* could penetrate the northern latitudes. Exactly what, if anything, these implements could accomplish that the simple Oldowan flakes, choppers and scrapers that preceded them could not is unknown, although perhaps they conferred a better means of butchering.

But the Dmanisi finds of primitive hominids and Oldowan-like industries raise once again the question of what prompted our ancestors to leave their natal land.

Yet, there is one major problem with scenarios involving departure dates earlier than about 1.7-1.4 million years ago, and that is simply that they involve geographic spread before the cultural developments (Acheulean industry, meat eating, fire, shelter) that are supposed to have made it possible.

A shift toward meat eating might explain how humans managed to survive outside of Africa, but what prompted them to push into new territories remains unknown at this time.

Perhaps they were following herds of animal north. Or maybe it was as simple and familiar as a need to know what lay beyond that hill or river or tall savanna grass. Also an early migration could explain technological differences between western and eastern Homo erectus populations.

Summary

Overall, the evidence suggests that *Homo ergaster* was the first hominid species to resemble historic hunter-gatherers not only in a fully terrestrial lifestyle, but also in a social organization that featured economic cooperation between males and females and perhaps between semi-permanent male-female units.

MIDDLE PALEOLITHIC HOMINIDS

The second phase of human migration

The time period between 250,000 and 50,000 years ago is commonly called the Middle Paleolithic.

At the same time that Neanderthals occupied Europe and Western Asia, other kinds of people lived in the Far East and Africa, and those in Africa were significantly more modern than the Neanderthals.

These Africans are thus more plausible ancestors for living humans, and it appears increasingly likely that Neanderthals were an evolutionary dead end, contributing few if any genes to historic populations.

Topics to be covered in this chapter:

- Summary of the fossil evidence for both the Neanderthals and some of their contemporaries;

- Second phase of human migration ("Out-of-Africa 2" Debate)

Neanderthals

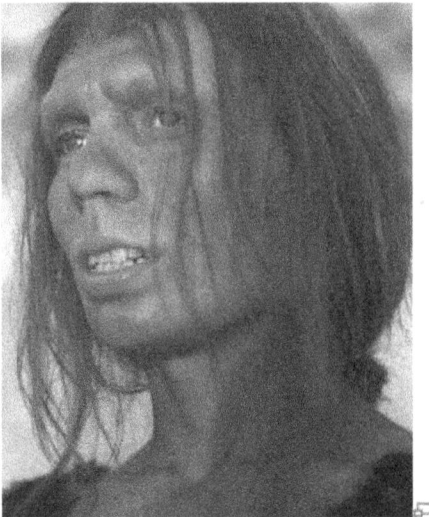

Museum recronstruction model of a Neanderthal woman

History of Research

In 1856, a strange skeleton was discovered in Feldhofer Cave in the Neander Valley ("thal" = valley) near Dusseldorf, Germany. The skull cap was as large as that of a present-day human but very different in shape. Initially this skeleton was interpreted as that of a congenital idiot.

The Forbes Quarry (Gibraltar) female cranium (now also considered as Neanderthal) was discovered in 1848, eight years before the Feldhofer find, but its distinctive features were not recognized at that time.

Subsequently, numerous Neanderthal remains were found in Belgium, Croatia, France, Spain, Italy, Israel and Central Asia.

Anthropologists have been debating for 150 years whether Neanderthals were a distinct species or an ancestor of Homo sapiens sapiens. In 1997, DNA analysis from the Feldhofer Cave specimen showed decisively that Neanderthals were a distinct lineage.

These data imply that Neanderthals and *Homo sapiens sapiens* were separate lineages with a common ancestor, Homo heidelbergensis, about 600,000 years ago.

Anatomy

Unlike earlier hominids (with some rare exceptions), Neanderthals are represented by many complete or nearly complete skeletons. Neanderthals provide the best hominid fossil record of the Plio-Pleistocene, with about 500 individuals. About half the skeletons were children. Typical cranial and dental features are present in the young individuals, indicating Neanderthal features were inherited, not acquired.

Morphologically the Neanderthals are a remarkably coherent group. Therefore they are easier to characterize than most earlier human types.

Neanderthal skull has a low forehead, prominent brow ridges and occipital bones. It is long and low, but relatively thin walled. The back of the skull has a characteristic rounded bulge, and does not come to a point at the back.

Cranial capacity is relatively large, ranging from 1,245 to 1,740 cc and averaging about 1,520 cc. It overlaps or even exceeds average for *Homo sapiens sapiens*. The robust face with a broad nasal region projects out from the braincase. By contrast, the face of modern Homo sapiens sapiens is tucked under the brain box, the forehead is high, the occipital region rounded, and the chin prominent.

Neanderthals have small back teeth (molars), but incisors are relatively large and show very heavy wear.

Neanderthal short legs and arms are characteristic of a body type that conserves heat. They were strong, rugged and built for cold weather. Large elbow, hip, knee joints, and robust bones suggest great muscularity. Pelvis had longer and thinner pubic bone than modern humans.

All adult skeletons exhibit some kind of disease or injury. Healed fractures and severe arthritis show that they had a hard life, and individuals rarely lived past 40 years old.

Chronology

Neanderthals lived from about 250,000 to 30,000 years ago in Eurasia.

The earlier ones, like at Atapuerca (Sima de Los Huesos), were more generalized. The later ones are the more specialized, "classic" Neanderthals.

The last Neanderthals lived in Southwest France, Portugal, Spain, Croatia, and the Caucasus as recently as 27,000 years ago.

Geography

The distribution of Neanderthals extended from Uzbekistan in the east to the Iberian peninsula in the west, from the margins of the Ice Age glaciers in the north to the shores of the Mediterranean sea in the south.

South-West France (Dordogne region) is among the richest in Neanderthal cave shelters:

- La Chapelle-aux-Saints;
- La Ferrassie;
- Saint-CÃ(c)saire (which is one of the younger sites at 36,000).

Other sites include:

- Krapina in Croatia;
- Saccopastore in Italy;
- Shanidar in Iraq;
- Teshik-Tash (Uzbekistan). The 9-year-old hominid from this site lies at the most easterly known part of their range.

No Neanderthal remains have been discovered in Africa or East Asia.

Homo Sapiens

Chronology and Geography

The time and place of Homo sapiens origin has preoccupied anthropologists for more than a century. For the longest time, many assumed their origin was in South-West Asia. But in 1987, anthropologist Rebecca Cann and colleagues compared DNA of Africans, Asians, Caucasians, Australians, and New Guineans. Their findings were striking in two respects:

- the variability observed within each population was greatest by far in Africans, which implied the African population was oldest and thus ancestral to the Asians and Caucasians;

- there was very little variability between populations which indicated that our species originated quite recently.

The human within-species variability was only 1/25th as much as the average difference between human and chimpanzee DNA. The human and chimpanzee lineages diverged about 5 milllion years ago. 1/25th of 5 million is 200,000. Cann therefore concluded that Homo sapiens originated in Africa about 200,000 years ago. Much additional molecular data and hominid remains further support a recent African origin of Homo sapiens, now estimated to be around 160,000-150,000 years ago.

Earliest Evidence

The fossil and archaeological finds characteristic of early modern humans are represented at various sites in East and South Africa, which date between 160,000 and 77,000 years ago.

Herto (Middle Awash, Ethiopia)

In June 2003, publication of hominid remains of a new subspecies: *Homo sapiens idaltu*. Three skulls (two adults, one juvenile) are interpreted as the earliest near-modern humans: 160,000-154,000 BP. They exhibit some modern traits (very large cranium; high, round skull; flat face without browridge), but also retain archaic features (heavy browridge; widely spaced eyes). Their anatomy and antiquity link earlier archaic African forms to later fully modern ones, providing strong evidence that East Africa was the birthplace of *Homo sapiens*.

Omo Kibish (Ethiopia)

In 1967, Richard Leakey and his team uncovered a partial hominid skeleton (Omo I), which had the features of Homo sapiens. Another partial fragment of a skull (Omo II) revealed a cranial capacity over 1,400 cc. Dating of shells from the same level gave a date of 130,000 years.

Ngaloba, Laetoli area (Tanzania)

A nearly complete skull (LH 18) was found in Upper Ngaloba Beds. Its morphology is largely modern, yet it retains some archaic features such as prominent brow ridges and a receding forehead. Dated at about 120,000 years ago.

Border Cave (South Africa)

Remains of four individuals (a partial cranium, 2 lower jaws, and a tiny buried infant) were found in a layer dated to at least 90,000 years ago. Although fragmentary, these fossils appeared modern.

Klasies River (South Africa)

Site occupied from 120,000 to 60,000 years ago. Most human fossils come from a layer dated to around 90,000 years ago. They are fragmentary: cranial, mandibular, and postcranial pieces. They appear modern, especially a fragmentary frontal bone that lacks a brow ridge. Chin and tooth size also have a modern aspect.

Blombos Cave (South Africa)

A layer dated to 77,000 BCE yielded 9 human teeth or dental fragments, representing five to seven individuals, of modern appearance.

Anatomy

African skulls have reduced browridges and small faces. They tend to be higher, more rounded than classic Neanderthal skulls, and some approach or equal modern skulls in basic vault shape. Where cranial capacity can be estimated, the African skulls range between 1,370 and 1,510 cc, comfortably within the range of both the Neanderthals and anatomically modern people.

Mandibles tend to have significantly shorter and flatter faces than did the Neanderthals.

Postcranial parts indicate people who were robust, particularly in their legs, but who were fully modern in form.

Out-of-Africa 2: The debate

Most anthropologists agree that a dramatic shift in hominid morphology occurred during the last glacial epoch. About 150,000 years ago the world was inhabited by a morphologically heterogeneous collection of hominids: Neanderthals in Europe; less robust archaic Homo sapiens in East Asia; and somewhat more modern humans in East Africa (Ethiopia) and also SW Asia. By 30,000 years ago, much of this diversity had disappeared. Anatomically modern humans occupied all of the Old World.

In order to understand how this transition occurred, we need to answer two questions:

1.   - Did the genes that give rise to modern human morphology arise in one region, or in many different parts of the globe?

2.   - Did the genes spread from one part of the world to another by gene flow, or through the movement and replacement of one group of people by another?

Unfortunately, genes don't fossilize, and we cannot study the genetic composition of ancient hominid populations directly. However, there is a considerable amount of evidence that we can bring to bear on these questions through the anatomical study of the fossil record and the molecular biology of living populations.

Two opposing hypotheses for the transition to modern humans have been promulgated over the last decades:

* the "multi-regional model" sees the process as a localized speciation event;

* the "out-of-Africa model" sees the process as the result of widespread phyletic transformation.

The "Multi-regional" model

This model proposes that ancestral Homo erectus populations throughout the world gradually and independently evolved first through archaic Homo sapiens, then to fully modern humans. In this case, the Neanderthals are seen as European versions of archaic sapiens.

Recent advocates of the model have emphasized the importance of gene flow among different geographic populations, making their move toward modernity not independent but tied together as a genetic network over large geographical regions and over long periods of time. Since these populations were separated by great distances and experienced different kinds of environmental conditions, there was considerable regional variation in morphology among them.

One consequence of this widespread phyletic transformation would be that modern geographic populations would have very deep genetic roots, having begun to separate from each other a very long time ago, perhaps as much as a million years.

This model essentially sees multiple origins of Homo sapiens, and no necessary migrations.

The "Out-of-Africa"/"Replacement" model

This second hypothesis considers a geographically discrete origin, followed by migration throughout the rest of the Old World. By contrast with the first hypothesis, here we have a single origin and extensive migration.

Modern geographic populations would have shallow genetic roots, having derived from a speciation event in relatively recent times. Hominid populations were genetically isolated from each other during the Middle Pleistocene. As a result, different populations of Homo erectus and archaic Homo sapiens evolved independently, perhaps forming several hominid species. Then, between 200,000 and 100,000 years ago, anatomically modern humans arose someplace in Africa and spread out, replacing other archaic sapiens including Neanderthals. The replacement model does not specify how anatomically modern humans usurped local populations. However, the model posits that there was little or no gene flow between hominid groups.

Hypothesis testing

If the "Multi-regional Model" were correct, then it should be possible to see in modern populations echoes of anatomical features that stretch way back into prehistory: this is known as regional continuity. In addition, the appearance in the fossil record of advanced humans might be expected to occur more or less simultaneously throughout the Old World. By contrast, the "Out-of-Africa Model" predicts little regional continuity and the appearance of modern humans in one locality before they spread into others.

Out-of-Africa 2: The evidence

Until relatively recently, there was a strong sentiment among anthropologists in favor of extensive regional continuity. In addition, Western Europe tended to dominate the discussions. Evidence has expanded considerably in recent years, and now includes molecular biology data as well as fossils. Now there is a distinct shift in favor of some version of the "Out-of-Africa

Model".

Discussion based on detailed examination of fossil record and mitochondrial DNA needs to address criteria for identifying:

- regional continuity;

- earliest geographical evidence (center of origin);

- chronology of appearance of modern humans.

Fossil record

Regional Continuity

The fossil evidence most immediately relevant to the origin of modern humans is to be found throughout Europe, Asia, Australasia, and Africa, and goes back in time as far as 300,000 years ago.

Most fossils are crania of varying degrees of incompleteness. They look like a mosaic of Homo erectus and Homo sapiens, and are generally termed archaic sapiens. It is among such fossils that signs of regional continuity are sought, being traced through to modern populations.

For example, some scholars (Alan Thorne) argue for such regional anatomical continuities among Australasian populations and among Chinese populations. In the same way, some others believe a good case can be made for regional continuity in Central Europe and perhaps North Africa.

Replacement

By contrast, proponents of a replacement model argue that, for most of the fossil record, the anatomical characters being cited as indicating regional continuity are primitive, and therefore cannot be used uniquely to link specific geographic populations through time.

The equatorial anatomy of the first modern humans in Europe presumably is a clue to their origin: Africa. There are sites from the north, east and south of the African continent with specimens of anatomical modernity. One of the most accepted is Klasies River in South Africa. The recent discovery of remains of H. sapiens idaltu at Herto (Ethiopia) confirms this evidence. Does this mean that modern Homo sapiens arose as a speciation event in Eastern Africa (Ethiopia), populations migrating north, eventually to enter Eurasia? This is a clear possibility.

The earlier appearance of anatomically moderns humans in Africa than in Europe and in Asia too supports the "Out-of-Africa Model".

Molecular biology

Just as molecular evidence had played a major role in understanding the beginnings of the hominid family, so too could it be applied to the later history, in principle.

However, because that later history inevitably covers a shorter period of time - no more than the past 1 million years - conventional genetic data would be less useful than they had been for pinpointing the time of divergence between hominids and apes, at least 5 million years ago. Genes in cell nuclei accumulate mutations rather slowly. Therefore trying to infer the recent history of populations based on such mutations is difficult, because of the relative paucity of information. DNA that accumulates mutations at a much higher rate would, however, provide adequate information for reading recent population history. That is precisely what mitochondrial DNA (mtDNA) offers.

MtDNA is a relatively new technique to reconstruct family trees. Unlike the DNA in the cell nucleus, mtDNA is located elsewhere in the cell, in compartments that produce the energy needed to keep cells alive. Unlike an individual's nuclear genes, which are a combination of genes from both parents, the mitochondrial genome comes only from the mother. Because of this maternal mode of inheritance, there is no recombination of maternal and paternal genes, which sometimes blurs the history of the genome as read by geneticists. Potentially, therefore, mtDNA offers a powerful way of inferring population history.

MtDNA can yield two major conclusions relevant for our topic: the first addresses the depth of our genetic routes, the second the possible location of the origin of anatomically modern humans.

Expectations

Multiregional model:

- extensive genetic variation, implying an ancient origin, going back at least a million years (certainly around 1.8 million years ago);

- no population would have significantly more variation than any other. Any extra variation the African population might have had as the home of *Homo erectus* would have been swamped by the subsequent million years of further mutation.

Replacement model:

- limited variation in modern mtDNA, implying a recent origin;

- African population would display most variation.

Results

1.   . If modern populations derive from a process of long regional continuity, then mtDNA should reflect the establishment of those local populations, after 1.8 million years ago, when populations of Homo erectus first left Africa and moved into the rest of the Old World. Yet the absence of ancient mtDNA in any modern living population gives a different picture. The amount of genetic variation throughout all modern human populations is surprisingly small, and implies therefore a recent origin for the common ancestor of us all.

2.   . Although genetic variation among the world's population is small overall, it is greatest in African populations, implying they are the longest established.

3.　　　. If modern humans really did evolve recently in Africa, and then move into the rest of the Old World where they mated with established archaic sapiens, the resulting population would contain a mixture of old and new mtDNA, with a bias toward the old because of the relative numbers of newcomers to archaic sapiens. Yet the evidence does not seem to support this view.

The argument that genetic variation among widely separated populations has been homogenized by gene flow (interbreeding) is not tenable any more, according to population geneticists.

Intermediate Model

Although these two hypotheses dominate the debate over the origins of modern humans, they represent extremes; and there is also room for several intermediate models.

- One hypothesis holds that there might have been a single geographic origin as predicted by replacement model, but followed by migrations in which newcomers interbred with locally established groups of archaic sapiens. Thus, some of genes of Neanderthals and archaic H. sapiens may still exist in modern populations;

- Another hypothesis suggests that there could have been more extensive gene flow between different geographic populations than is allowed for in the multi-regional model, producing closer genetic continuity between populations. Anatomically modern humans evolved in Africa, and then their genes diffused to the rest of the world by gene flow, not by migration of anatomically modern humans and replacement of local peoples.

In any case the result would be a much less clearcut signal in the fossil record.

Case studies

Southwest Asia

Neanderthal fossils have been found in Israel at several sites: Kebara, Tabun, and Amud. For many years there were no reliable absolute dates. Recently, these sites were securely dated. The Neanderthals occupied Tabun around 110,000 years ago. However, the Neanderthals at Kebara and Amud lived 55,000 to 60,000 years ago. By contrast, at Qafzeh Cave, located nearby, remains currently interpreted as of anatomically modern humans have been found in a layer dated to 90,000 years ago.

These new dates lead to the surprising conclusion that Neanderthals and anatomically modern humans overlapped - if not directly coexisted - in this part of the world for a very long time (at least 30,000 years). Yet the anatomical evidence of the Qafzeh hominid skeletons reveals features reminiscent of Neanderthals. Although their faces and bodies are large and heavily built by today's standards, they are nonetheless claimed to be within the range of living peoples. Yet, a recent statistical study comparing a number of measurements among Qafzeh, Upper Paleolithic and Neanderthal skulls found those from Qafzeh to fall in between the Upper Paleolithic and Neanderthal norms, though slightly closer to the Neanderthals.

Portugal

The Lagar Velho 1 remains, found in a rockshelter in Portugal dated to 24,500 years ago, correspond to the complete skeleton of a four-year-old child.

This skeleton has anatomical features characteristic of early modern Europeans:

- prominent chin and certain other details of the mandible;

- small front teeth;

- characteristic proportions and muscle markings on the thumb;

- narrowness of the front of pelvis;

- several aspects of shoulder and forearm bones.

Yet, intriguingly, a number of features also suggest Neanderthal affinities:

- the front of the mandible which slopes backward despite the chin;

- details of the incisor teeth;

- pectoral muscle markings;

- knee proportions and short, strong lower-leg bones.

Thus, the Lagar Velho child appears to exhibit a complex mosaic of Neanderthal and early modern human features. This combination can only have resulted from a mixed ancestry; something that had not been previously documented for Western Europe. The Lagar Velho child is interpreted as the result of interbreeding between indigenous Iberian Neanderthals and early modern humans dispersing throughout Iberia sometime after 30,000 years ago. Because the child lived several millennia after Neanderthals were thought to have disappeared, its anatomy probably reflects a true mixing of these populations during the period when they coexisted and not a rare chance mating between a Neanderthal and an early modern human.

Population dispersal into Australia/Oceania

Based on current data (and conventional view), the evidence for the earliest colonization of Australia would be as follows:

- archaeologists have generally agreed that modern humans arrived on Australia and its continental islands, New Guinea and Tasmania, about 35,000 to 40,000 years ago, a time range that is consistent with evidence of their appearance elsewhere in the Old World well outside Africa;

- all hominids known from Greater Australia are anatomically modern Homo sapiens;

- emerging picture begins to suggest purposeful voyaging by groups possessed of surprisingly sophisticated boat-building and navigation skills;

- the only major feature of early Greater Australia archaeology that does NOT fit comfortably with a consensus model of modern human population expansion in the mid-Upper Pleistocene is the lithic technology, which has a pronounced Middle, rather than Upper, Paleolithic cast.

Over the past decade, however, this consensus has been eroded by the discovery and dating of several sites:

- Malakunanja II and Nauwalabila I, located in Arnhem Land, would be 50,000 to 60,000 years old;

- Jinmium yielded dates of 116,000 to 176,000 years ago.

Yet these early dates reveal numerous problems related to stratigraphic considerations and dating methods. Therefore, many scholars are skeptical of their value.

If accurate, these dates require significant changes in current ideas, not just about the initial colonization of Australia, but about the entire chronology of human evolution in the early Upper Pleistocene. Either fully modern humans were present well outside Africa at a surprisingly early date or the behavioral capabilities long thought to be uniquely theirs were also associated, at least to some degree, with other hominids.

As a major challenge, the journey from Southeast Asia and Indonesia to Australia, Tasmania and New Guinea would have required sea voyages, even with sea levels at their lowest during glacial maxima. So far, there is no archaeological evidence from Australian sites of vessels that could have made such a journey. However, what were coastal sites during the Ice Age are mostly now submerged beneath the sea.

Summary

Overall the evidence suggested by mitochondrial DNA is the following:

- the amount of genetic variation in human mitochondrial DNA is small and implies a recent origin for modern humans;

- the African population displays the greatest amount of variation; this too is most reasonably interpreted as suggesting an African origin.

MIDDLE PALEOLITHIC TOOL AND SUBSISTENCE PATTERNS

Stone tool industry

Neanderthals and their contemporaries seem to have been associated everywhere with similar stone tool industries, called the Mousterian (after Le Moustier Cave in France). Therefore no fundamental behavioral difference is noticeable. The implication may be that the anatomical differences between Neanderthals and near-moderns have more to do with climatic adaptation and genetic flow than with differences in behavior.

Archaeological sites are dominated by flake tools. By contrast, Acheulean sites are dominated by large handaxes and choppers. Handaxes are almost absent from Middle Paleolithic sites. Oldowan hominids used mainly flake tools as well. However, unlike the small, irregular Oldowan flakes, the Middle Paleolithic hominids produced quite symmetric, regular flakes using sophisticated methods.

The main method is called the Levallois and it involves three steps in the core reduction:

1.   - the flintknapper prepares a core having one precisely shaped convex surface;

2.   - then, the knapper makes a striking platform at one end of the core;

3.   - finally, the knapper hits the striking platform, knocking off a flake whose shape is determined by the original shape of the core.

Mousterian tools are more variable than Acheulean tools. Traditionally tools have been classified into a large number of distinct types based on their shape and inferred function. Among the most important ones are:

* points;

* side scrapers, flake tools bearing a retouched edge on one or both sides;

* denticulates, flake tools with a succession of irregular adjacent notches on one or both sides.

Francois Bordes found that Middle Paleolithic sites did not reveal a random mix of tool types, but fell into one of four categories that he called Mousterian variants. Each variant had a different mix of tool types. Bordes concluded that these sites were the remains of four wandering groups of Neanderthals, each preserving a distinct tool tradition over time and structured much like modern ethnic groups.

Recent studies give reason to doubt Bordes' interpretation. Many archaeologists believe that the variation among sites results from differences in the kinds of activity performed at each locality. For example, Lewis Binford argued that differences in tool types depend on the nature of the site and the nature of the work performed. Some sites may have been base camps where people lived, while others may have been camps at which people performed subsistence tasks. Different tools may have been used at different sites for woodworking, hide preparation, or butchering prey.

Recently, however, microscopic studies of wear patterns on Mousterian tools suggest that the majority of tools were used mainly for woodworking. As a result, there seems to be no association between a tool type (such as a point or a side-scraper) and the task for which it was used.

Microscopic analyses of the wear patterns on Mousterian tools also suggest that stone tools were hafted, probably to make spears.

Mousterian hominids usually made tools from rocks acquired locally. Raw materials used to make tools can typically be found within a few kilometers of the site considered.

Subsistence Patterns

Neanderthal sites contain bones of many animals alongside Mousterian stone tools. European sites are rich in bones of red deer, fallow deer, bison, aurochs, wild sheep, wild goat and horse, while eland, wildebeest, zebra are found often at African sites. Archaeologists find only few bones of very large animals such as hippopotamus, rhinoceros and elephant, even though they were plentiful in Africa and Europe.

This pattern has provoked as much debate at similar ambiguity as for earlier hominids, regarding the type of food procurement responsible for the presence of these bones: hunting or scavenging.

Several general models have be offered to explain the Mousterian faunal exploitation:

Obligate Scavenger Model

Some archaeologists (such as Lewis Binford) believe that Neanderthals and their contemporaries in Africa never hunted anything larger than small antelope, and even these prey were acquired opportunistically, not as a result of planned hunts. Any bones of larger animals were acquired by scavenging. As evidence in support of this view, the body parts which dominate (skulls and foot bones) are commonly available to scavengers. Binford believes that hominids of this period did not have the cognitive skills necessary to plan and organize the cooperative hunts necessary to bring down large prey. Mousterian hominids were nearly as behaviorally primitive as early Homo.

Flexible Hunter-Scavenger Model

Other scientists argue that Neanderthals likely were not obligate scavengers, but that during times when plant foods were abundant they tended to scavenger rather than hunt. At other times, when plant foods were less abundant, Neanderthals hunted regularly. Their interpretation is of a flexible faunal exploitation strategy that shifted between hunting and scavenging.

Less-Adept Hunter Model

Other scientists believe that Neanderthals were primarily hunters who regularly killed large animals. But they were less effective hunters than are modern humans. They point out that animal remains at Mousterian sites are often made up of one or two species:

For example, at Klasies River, vast majority of bones are from eland.

The prey animals are large creatures, and they are heavily overrepresented at these sites compared with their occurrence in the local ecosystem. It is hard to see how an opportunistic scavenger would acquire such a non-random sample of the local fauna. One important feature of this model is that animal prey hunted such as eland are not as dangerous prey as buffalo. Middle Paleolithic hominids were forced to focus on the less dangerous (but less abundant) eland, because they were unable to kill the fierce buffalo regularly.

Fully Adept Hunter Model

Finally some scientists argue that scavenging was not a major component of the Middle Paleolithic diet and there is little evidence of a less effective hunting strategy. Skeletal element abundance and cut mark patterning would be consistent with hunting.

Overall, there is currently no evidence that Middle Paleolithic hominids differed from Upper Paleolithic hominids in scavenging or hunting, the most fundamental aspect of faunal exploitation. The differences that separate Middle Paleolithic hominids from modern hominids may not reside in scavenging versus hunting or the types of animals that they pursued. Differences in the effectiveness of carcass use and processing, with their direct implications for caloric yield, may be more important.

Neanderthals lacked sophisticated meat and fat storage technology, as well as productive fat rendering technology. At a minimum, the lack of storage capabilities and a lower caloric yield per carcass have forced Neanderthals to use larger foraging ranges to increase the likelihood of successful encounters with prey.

Cannibalism

Marks on human bones from Middle Paleolithic can be the result of two phenomena: violence and cannibalism.

Violence

Violence can be recognized on bone assemblages by:

- marks of weapons;
- cutmarks on skull produced by scalping;
- removal of heads and hands as trophies;
- breakage of faces;
- much less "body processing" than in case of cannibalism.

Evidence for violence in the Middle Paleolithic is extremely rare.

Body processing and cannibalism

By contrast, evidence of body processing and cannibalism is becoming more widespread at different times and in different geographical areas.

Chronology

Lower Paleolithic

Sterkfontein (South Africa): cannibalism

Bodo cranium (Ethiopia): cannibalism

Atapuerca (Spain): cannibalism

Middle Paleolithic (Neanderthals)

Krapina (Croatia): body processing

Moula-Guercy (France): cannibalism

Marillac (France): cannibalism

Combe-Grenal (France): cannibalism

Middle Paleolithic (Homo sapiens idaltu)

Herto (Ethiopia): body processing

Upper Paleolithic (with Neanderthals)

Vindija (Croatia): cannibalism

Zafarraya (Spain): cannibalism

Neolithic

FontbrÃ(c)goua (France): cannibalism

Criteria

Criteria required for a "minimal taphonomic signature" of cannibalism:

- breakage of bones (to get at marrow and brain);
- cut marks suggesting butchery;
- so-called anvil abrasions left where a bone has rested on a stone anvil whilst it is broken with a hammer stone;
- burning (after breakage and cutting);

- virtual absence of vertebrae (crushed or boiled to get at marrow and grease);

- "pot polishing" on the ends of bones which have been cooked and stirred in a clay pot.

  These criteria **must** be found on **both** hominid and ungulate remains. Finally the types of bones usually broken are the crania and limb bones.

Patterns

Different behavioral patterns toward the dead among Middle Paleolithic Neanderthals:

- Cannibalism: Moula-Guercy

- Human individuals defleshed and disarticulated.

- Bones smashed for marrow and brain.

- Mortuary practices with body processing:Krapina, Herto

- Postmortem processing of corpses with stone tools, probably in preparation for burial of cleaned bones.

- No evidence of marrow processing.

- Mortuary practices without body processing: Southwest Asia (Amud, Kebara, Shanidar)

- Intentional burials; dead bodies placed carefully in burial pits with tools and grave goods.

EARLY UPPER PALEOLITHIC CULTURES

Aurignacian

First Discovered

- Aurignac (Dordogne, France)

Chronology

- ca. 35,000-27,000 BCE

Geography

- Widespread distribution over Eurasia

Hominid

- Modern humans (Homo sapiens)

Material Culture

- Upper Paleolithic-type lithic industry
- Aurignacian blades, burins, endscrapers, etc.
- Bone Tools

Mortuary practices

- Definitive elaborate burials, with grave goods

Symbolic Expression

Proliferation of various forms of personal ornaments:

- perforated animal-teeth;
- distinctive "bead" forms carved out of bone and mammoth ivory;
- earliest perforated marine shells

Artistic Expression

Types of evidence:

- Engraved limestone blocks
- Animal and human figurines

- Parietal art

Engraved block characteristics:

- Stiffness of outlines;

- Deep incisions;

- Work executed mainly on limestone slabs or blocks;

- Sexual symbols realistically represented;

- Animals (usually heads, forequarters and dorsal lines) extremely crudely rendered;

- This type of artistic expression limited to southwest France (mainly Dordogne).

Figurine characteristics:

- Earliest evidence of artwork in the Upper Paleolithic: Geissenklösterle - 37,000-33,000 BCE

- Present in Central Europe, presently Germany

- Sophisticated and naturalistic statuettes of animal (mammoth, feline, bear, bison) and even human figures

- Carved from mammoth ivory

Gravetian

First Discovered

- La Gravette (Dordogne, France)

Chronology

- ca. 27,000-21,000 BCE

Geography

- Widespread distribution over Eurasia

Major cultural centers

- Southwest France

- Northern Italy (Grimaldi)

- Central Europe (Dolni Vestonice, Pavlov)

- Siberia (Sungir)

Architecture

- Mammoth huts

Material Culture

- Upper Paleolithic-type lithic industry
- Gravette Points, etc.

Other Economic Activities

- Pyrotechnology
- Basketry

Complex mortuary practices

- Dolni Vestonice triple burial

Artisitic Expression

Types:

- Animal figurines
- Female figurines ("Venuses")
- Parietal art

Animal figuring characteristics: Animals most frequently depicted are dangerous species (felines and bears), continuing Aurignacian tradition

- In Moravia, 67 animal statuettes recorded:
- 21 bears
- 11 small carnivores
- 9 felines
- 8 mammoths
- 6 birds
- 6 horses

- 4 rhinoceroses

- caprid

- 1 cervid

  By contrast, Magdalenian animal statuettes from the same region show very different patterns (N=139):

- 56 horses, 44 bisons

- 9 bears,

- 2 felines,

- 1 mammoth

- 2 birds

- 1 caprid, 1 cervid

- 5 miscellaneous, 18 indeterminates

- No rhinoceros

  Dangerous animals represent only 10% of total

  Female figurrine characteristics: Widespread distribution over Europe and Russia; except Spain where no evidence of Venuses

- Raw materials:

- ivory

- clay

- Various types of research performed by anthropologists:

- technological

- stylistic

- details of clothing, ornaments

- chronological/geographical

- interpretational

- Most of baked clay figurines found fragmented

- Lack of skill or deliberate action? Intentional fracturation through heating process
- Fragmented figurines were intended products Involved and by-products of ritual ceremonies rather than art objects

  Parietal art characteristics: From 21 sites, a list of 47 animals identified:

- 9 ibexes
- 9 cervids
- 7 horses
- 4 mammoths
- 3 bovids
- 1 megaceros
- 1 salmon
- 10 indeterminates

  Dangerous animals (rhinoceros, bear, lion) depicted during the Gravettian do not constitute more than 11% of determinable animals:

- 3 times less than in Aurignacian period);
- yet still higher frequency than during Solutrean and Magdalenian

  Strong preponderance of hunted animals, with horse very widely dominant

- Example: Gargas with a list of 148 animals identified:
- 36.5% bovids (bison and aurochs)
- 29% horses
- 10% ibexes
- 6% cervids
- 4% mammoths
- 8% indeterminates
- (2 birds, 1 wild boar)
- **No** feline, rhinoceros, bear

LATE UPPER PALEOLITHIC CULTURES

Solutrean

First Discovered

- SolutrÃ(c) (NE France)

Chronology

- ca. 21,000-18,000 BCE

Geography

- Limited distribution over SW France and Iberia

Material Culture

- Upper Paleolithic-type lithic industry
- Heat Treatment, Pressure Retouch
- Solutrean points: bifacially retouched leaf points, shouldered points, etc.
- burins, endscrapers, etc.

Settlements

- Some sedentary groups (Fourneau-du-Diable)
- Long stratigraphic sequences

Human remains

- Complex mortuary practices:

  No evidence of burials, but manipulation of dead (e.g., reuse of skull: Le Placard)

Artistic expression

Types:

- Engraved limestone blocks
- Engraved Bones
- Parietal art

Characteristics:

- Various techniques applied: painting, engraving
- Distribution and amount of animals represented in tradition of Late Upper Paleolithic: mainly horses and bisons
- Several novelties from Gravettian:
- First association of parietal art with occupation sites [Low-relief scupture on blocks detached from walls];
- Representation of animals in line or opposed

Magdelenian

First Discovered

- La Madeleine (Dordogne, France)

Chronology

- ca. 19,000-10,000 BCE

Geography

- Widespread distribution over Eurasia

Major cultural centers

- Southwest France (Charente, Dordogne, PyrÃ(c)nÃ(c)es)
- Northeast Spain
- Central Europe
- Siberia

Material Culture

- Upper Paleolithic-type lithic industry
- Magdalenian blades, burins, etc.
- Rich bone tool industry (harpoons)

Complex mortuary practices

- Children burials

Artistic expression

Types:

- Raw Materials: Great diversity (limestone cave walls and slabs, sandstone, shale, bone, ivory, clay, etc.)

- Techniques: All techniques employed: Engraving, Sculpture, Molding, Cutting, Drawing, Painting

- Both mobiliary and parietal arts present. Out of about 300 sites with parietal art, 250 are attributed to Magdalenian period.

- Types of Figurations:

- Animals (mainly horses and bisons)

- Humans (male and female)

- Hands (positive and negative)

- Signs (dots, lines

CHRONOLOGY AND DATING TECHNIQUES

Having an accurate time scale is a crucial aspect of reconstructing how anatomical and behavioral characteristics of early hominids evolved.

Researchers who are interested in knowing the age of particular hominid fossils and/or artifacts have several options that fall into three areas:

- Stratigraphy

- Relative dating techniques

- Absolute dating techniques

Principles of stratigraphy

This is the study of (sometimes) horizontal sedimentation or the accumulation of layers of earth sediments (called strata). Most paleoanthropology studies the earliest-known hominids in East Africa, which has a very specific geological and tectonic context where layers crucial for the study of human evolution are exposed.

Relative dating techniques

Relative dating techniques follow the same principle as "guilt by association." If a hominid bone is found alongside the bones of an animal that became extinct 1 million years ago, that may be evidence the hominid bone is at least 1 million years old. Objects used in relative dating include:

- Stone tools

- Faunal correlation

- Paleomagnetism

Absolute dating techniques

The majority of absolute dating techniques are radiometric, which means they exploit some aspect of measuring radioactive decay. This involves two principles. First, some action (heating) sets the radioactive "clock" to zero. Second, once the clock has been set to zero, the consequences of some kind of radioactive decay steadily accumulate, recording the passage of time. Materials measured in this process include:

- Potassium/argon

- Thermoluminesce/electron spin resonance

- Radiocarbon

METHODS OF DATING IN ARCHAEOLOGY

Techniques of recovery include:

- Surveys

- Excavations

Types of archaeological remains include:

- Perishable: plant remains, animal bones, wooden artifacts, basketry, and other easily degradable objects

- Nonperishable materials: stone tools, pottery, rocks used for structures.

Data collection and analysis is oriented to answer questions of subsistence, mobility or settlement patterns, and economy.

METHODS IN PHYSICAL ANTHROPOLOGY

Data collections based on study of hard tissues (bones and teeth), usually the only remains left of earlier populations, which include:

- Identification of bones/Which part of skeleton is represented?

- Measurement of the cranium and other elements of a skeleton. Carefully defined landmarks are established on the cranium, as well as on the long bones, to facilitate standardization of measurements

- Superficial examination of bone for any marks (for instance, cutmarks)

- Further examination using specific techniques:

- X-ray to identify evidence of disease and trauma in bones

- DNA extraction to determine genetic affiliations

CULTURAL EVOLUTION AND PROGRESS

The concept of progress

Progress is defined as a gradual but predictable bettering of the human condition from age to age.

History of Progressivism

- Progressivism has been one of the cornerstones of Western historical and philosophical thought since ancient Greek times.

- For most of its history (from the Greek period to the 15th century), Progressivism was a purely philosophical or ideological doctrine: in the absence of any empirical evidence of improvement in the human condition, early progressivists devised imaginary scenarios of human history in which they pictured the gradual emergence of enlightened present-day institutions out of earlier and ruder institutions. The defining characteristics of any primitive society, thus conceived, were wholly negative: they were whatever civilization was not.

- In the 15th century, ethnographic information about living "primitives," especially in the Americas, increased, providing the progressivists with empirical confirmation for their ideas. Living conditions of these "primitive peoples" conformed in a general way to the imagined early condition of humankind; they were therefore considered as "fossilized" survivals from ancient times, who had somehow failed to progress.

- In the 19th century, archaeology also began to provide confirmation: evidence of early peoples who had indeed lived without metals, agriculture, or permanent dwellings, just as the progressivists had always imagined.

- Out of the conjunction of progressivist theory with ethnographic and archaeological researches, the discipline of anthropology was born in the latter half of the 19th century. Early anthropologists arranged that evidence in orderly progression to provide seemingly irrefutable confirmation for what had long been only a philosophical doctrine. Progressivism was transformed into a science named anthropology; "Progress" was renamed "Social Evolution" or "Cultural Evolution."

Characteristics

Progressivism flourished mainly in optimistic times, that is, times of scientific advances and expanding imaginations:

- pre-Socratic Greece

- the Enlightenment (18th century)

- the Victorian age (19th century)

- the generation following World War II

Progressivism has been the doctrine that legitimizes all scientific discoveries and labels them as "advances".

- All progressivists agree, however, that their age is superior to those that preceded it.

- Human history is perceived with a basic directionality, from worse to better.

What is better?

- What constitutes "the better"? What causes it? How can it be measured?

- Answers to these questions have changed in accordance with the ideological preoccupation of different eras, and different philosophies:

- improved material circumstances

- intellectual maturation

- aesthetic achievements

Us vs. Them mentality

The problem with categorizing "progressive" judgments must be viewed in long-term perspective as a struggle between two basically incompatible cultural systems:

- in the present, us and the others: states/civilizations vs. bands/tribes;

- in paleoanthropology, us (*Homo sapiens*) and the others (other hominid species, e.g. Neandertals, *Homo erectus*).

Historical background of Western nations

Hunting and gathering was predominant as a way of life for about 7 million years, whereas life in cities or states has been around for only the past 5,000 years or so.

Changes, or progess, since the first appearance of urban life and state organization (5,000 yrs ago):

- non-state tribal peoples persisted in a dynamic equilibrium or symbiotic relationship with states/civilizations

- states/civilizations developed and remained within their own ecological boundaries

- this situation lasted for thousands of years

This situation shifted 500 years ago:

- In the 15th century, Europeans began to expand beyond their long-established frontiers.

- For about 250 years (until 1750), the expansion was relatively slow, as non-state tribal peoples still seemed secure and successfully adapted to their economically "marginal" refuges.

- In the mid-eighteenth century, the industrial revolution launched the developing Western nations on an explosive growth in population and consumption called "progress."

The Industrial Revolution

This period marks a major explosion at the scale of humankind:

- phenomenal increase in population

- increase in per capita consumption rates

- totally unparalleled scope

- these two critical correlates (population and consumption rates) of industrialization quickly led to overwhelming pressure on natural resources

Very quickly, industrial nations could no longer supply from within their own boundaries the resources needed to support further growth or even to maintain current consumption levels.

As a consequence:

- Industrial revolution led to an unprecedented assault on the world's relatively stable non-Western tribal peoples and their resources.

- Many of the "underdeveloped" resources controlled by the world's self-sufficient tribal peoples were quickly appropriated by outsiders to support their own industrial progress.

- In the last 200 years, these tribal cultures have virtually disappeared or have been completely marginalized.

Increased rates of resource consumption, accompanying industrialization, have been even more critical than mere population increase:

- Industrial civilization is a culture of consumption. In this respect, it differs most strikingly from tribal cultures.

- Industrial economies are founded on the principle that consumption must be ever expanded.

- Complex systems of mass marketing and advertising have been developed for that specific purpose.

- Social stratification in industrial societies is based primarily on inequalities in material wealth and is both supported and reflected by differential access to resources.

Industrial ideological systems and prejudices:

- place great stress on belief in
- continual economic growth
- progress
- measure "standard of living" in terms of levels of material consumption.

Ethnocentrism

Ethnocentrism is the belief in the superiority of one's own culture. It is vital to the integrity of any culture, but it can be a threat to the well-being of other peoples when it becomes the basis for forcing Western standards upon non-Western tribal cultures.

The impact of modern civilization on tribal peoples is a dominant research theme in anthropology and social sciences.

Among economic development writers, the consensus is the clearly ethnocentric view that any contact with superior industrial culture causes non-Western tribal peoples to voluntarily reject their own cultures in order to obtain a better life.

In the past, anthropologists also often viewed this contact from the same ethnocentric premises accepted by government officials, developers, missionaries, and the general public. But in recent years, there has been considerable confusion in the enormous culture change literature regarding the basic question of why tribal cultures seem inevitably to be acculturated or modernized by industrial civilization.

- There is therefore a problem to conceptualize the causes of the transformation process in simple nonethnocentric terms.

- This apparent inability may be due to the fact that the analysts are members of the culture of consumption that today happens to be the dominant world culture type.

- The most powerful cultures have always assumed a natural right to exploit the world's resources wherever they find them, regardless of the prior claims of indigenous populations.

Arguing for efficiency and survival of the fittest, old-fashioned colonialists elevated this "right" to the level of an ethical and legal principle that could be invoked to justify the elimination of any cultures that were not making "effective" use of their resources.

This viewpoint has found its way into modern theories of cultural evolution, expressed as the "Law of Cultural Dominance": *any cultural system which exploits more effectively the energy resources of a given environment will tend to spread in that environment at the expense of other less effective (indigenous) systems.*

- These old attitudes of social Darwinism are deeply embedded in our ideological system.

- They still occur in the professional literature on culture change.

While resource exploitation is clearly the basic cause of the destruction of tribal peoples, it is important to identify the underlying ethnocentric attitudes that are often used to justify what are actually exploitative policies.

Apart from the obvious ethical implications involved here, upon close inspection all of these theories expounding the greater adaptability, efficiency, and survival value of the dominant industrial culture prove to be quite misleading.

Of course, as a culture of consumption, industrial civilization is uniquely capable of consuming resources at tremendous rates, but this certainly does not make it a more effective culture than low-energy tribal cultures, if stability or long-run ecological success is taken as the criterion for "effectiveness."

Likewise, we should expect, almost by definition, that members of the culture of consumption would probably consider another culture's resources to be underexploited and to use this as a justification for appropriating them.

Among some writers, it is assumed that all people share our desire for what we define as material wealth, prosperity, and progress and that others have different cultures only because they have not yet been exposed to the superior technological alternatives offered by industrial civilization. Supporters of this view seem to minimize the difficulties of creating new wants in a culture and at the same time make the following highly questionable and clearly ethnocentric assumptions:

- The materialistic values of industrial civilization are cultural universals.

- Tribal cultures are unable to satisfy the material needs of their peoples.

- Industrial goods are, in fact, always superior to their handcrafted counterparts.

Assumption 1 - Unquestionably, tribal cultures represent a clear rejection of the materialistic values of industrial civilization, yet tribal individuals can indeed be made to reject their traditional values if outside interests create the necessary conditions for this rejection. The point is that far more is involved here than a mere demonstration of the superiority of industrial civilization.

Assumption 2 - The ethnocentrism of the second assumption is obvious. Clearly, tribal cultures could not have survived for millions of years if they did not do a reasonable job of satisfying basic human needs.

Assumption 3 - Regarding the third assumption, there is abundant evidence that many of the material accoutrements of industrial civilization may well not be worth their real costs regardless of how appealing they may seem in the short term.

PRE-DARWINIAN AND DARWINIAN THOUGHTS ON EVOLUTION

Pre-Darwinian Thoughts on Evolution

Throughout the Middle Ages, there was one predominant component of the European world view: stasis.

- All aspects of nature were considered as fixed and change was unconceivable.

- No new species had appeared, and none had disappeared or become extinct.

The social and political context of the Middle Ages helps explain this world view:

- shaped by feudal society - hierarchical arrangement supporting a rigid class system that had changed little for centuries

- shaped by a powerful religious system - life on Earth had been created by God exactly as it existed in the present (known as fixity of species).

This social and political context, and its world view, provided a formidable obstacle to the development of evolutionary theory. In order to formulate new evolutionary principles, scientists needed to:

- overcome the concept of fixity of species

- establish a theory of long geological time

From the 16th to the 18th century, along with renewed interest in scientific knowledge, scholars focused on listing and describing all kinds of forms of organic life. As attempts in this direction were made, they became increasingly impressed with the amount of biological diversity that confronted them.

These scholars included:

- John Ray (1627-1705) - put some order into the diversity of animal and plant life, by creating the concepts of species and genus.

- Carolus Linnaeus (1707-1778) - added two more categories (class and order) and created a complex system of classification (taxonomy) still used today; also innovated by including humans in his classification of animals.

- Georges-Louis Leclerc, Comte de Buffon (1707-1788) - innovated by suggesting the changing nature of species, through adaptation to local climatic and environmental conditions.

- Jean-Baptiste Lamarck (1744-1829) - offered a comprehensive system to explain species changes; postulated that physical alterations of organic life would occur in relation to changing environmental circumstances, making species better suited for their new habitat; also postulated that new traits would be passed on to offspring (the theory known as inheritance of acquired characteristics).

Therefore, the principle of "fixity of species" that ruled during the Middle Ages was no longer considered valid.

In the mid-19th century, Charles Darwin offered a new theory which pushed further the debate of evolutionary processes and marks a fundamental step in their explanation by suggesting that evolution works through natural selection.

Charles Darwin (1809-1882)

Charles Darwin's life as a scientist began when he took a position as naturalist aboard *HMS Beagle*, a ship charting the coastal waters of South America. As the ship circled the globe over a five-year period (1831-1836), Darwin puzzled over the diversity and distribution of life he observed. Observations and collections of materials made during these travels laid the foundation for his life's work studying the natural world.

As an example, the *Beagle* stopped five weeks in the Galapagos archipelago. There Darwin observed an unusual combination of species and wondered how they ended up on this island.

Darwin's observations on the diversity of plants and animals and their particular geographical distribution around the globe led him to question the assumption that species were immutable, established by a single act of creation. He reasoned that species, like the Earth itself, were constantly changing. Life forms colonized new habitats and had to survive in new conditions. Over generations, they underwent transmutation into new forms. Many became extinct. The idea of evolution slowly began to take shape in his mind.

In his 1859 publication *On the Origin of Species*, Darwin presented some of the main principles that explained the diversity of plants and animals around the globe: adaptation and natural selection. According to him, species were mutable, not fixed; and they evolved from other species through the mechanism of natural selection.

Darwin's theory of natural selection

In 1838, Darwin, at 28, had been back from his voyage on the *Beagle* for two years. He read Thomas Malthus's *Essay on Population*, which stated that human populations invariably grow until they are limited by starvation, poverty, and death, and realized that Malthus's logic could also apply to the natural world. This realization led Darwin to develop the principle of evolution by natural selection, which revolutionized our understanding of the living world.

His theory was published for the first time in 1859 in *On the Origin of Species by Means of Natural Selection, or the Preservation of Favoured Races in the Struggle for Life.*

Darwin's Postulates

The theory of adaptation and how species change through time follows three postulates:

- **Struggle for existence**: The ability of a population to expand is infinite, but the ability of any environment to support populations is always finite.

    Example: Animals require food to grow and reproduce. When food is plentiful, animal populations grow until their numbers exceed the local food supply. Since resources are always finite, it follows that not all individuals in a population will be able to survive and reproduce.

- **Variation in fitness**: Organisms in populations vary. Therefore, some individuals will possess traits that enable them to survive and reproduce more successfully (producing more offspring) than others in the same environment.

- **Inheritance of variation**: If the advantageous traits are inherited by offspring, then these traits will become more common in succeeding generations. Thus, traits that confer advantages in survival and reproduction are retained in the population, and traits that are disadvantageous disappear.

Examples of adaptation by natural selection

During his voyage on the *HMS Beagle*, Darwin observed a curious pattern of adaptations among several species of finches (now called Darwin's finches) that live on the Galapagos Islands.

Several traits of finches went through drastic changes in response to changes in their environment. One example is beak depth:

- There was huge variation in beak depth among finches on the island; it affected the birds' survival and adaptation to local environmental changes.

    During a drought, finches with deeper beaks were more likely to survive than finches with shallow beaks (which were at a disadvantage because it was harder for them to crack larger and harder seeds).

- Parents and offsprings had similar beak depths.

    Through natural selection, average morphology (an organism's size, shape and composition) of the bird population changed so that birds became better adapted to their environment.

Benefits and disadvantages of evolution

Individual Selection

Adaptation results from competition among individuals, not between entire populations or species.

Selection produces adaptations that benefit individuals. Such adaptation may or may not benefit the population or species. In the case of finches' beak depth, selection probably does allow the population of finches to compete more effectively with other populations of seed predators. However, this need not be the case. Selection often leads to changes in behavior or morphology that increase the reproductive success of individuals but decrease the average reproductive success and competitive ability of the group, population, and species.

Example of conflict between individual and group interests: All organisms in the population produce many more offspring than are necessary to maintain the species. A female monkey may, on average, produce 10 offspring during her lifetime. In a stable population, perhaps only two of these offspring will survive and reproduce. From the point of view of the species, the other eight are a waste of resources. The species as a whole might be more likely to survive if all females produced fewer offspring.

The idea that natural selection operates at the level of the individual is a key element in understanding adaptation.

Directional Selection

Instead of a completely random selection of individuals whose traits will be passed on to the next generation, there is selection by forces of nature. In this process, the frequency of genetic variants for harmful or maladaptive traits within the population is reduced while the frequency of genetic variants for adaptive traits is increased.

Natural selection, as it acts to promote change in gene frequencies, is referred to as directional selection.

Stabilizing Selection

Finches' beaks (Example)

Large beaks have benefits as well as disadvantages. Birds with large beaks are less likely to survive their juvenile period than birds with small beaks, probably because they require more food to grow.

Evolutionary theory prediction:

- Over time, selection will increase the average beak depth in a population until the costs of larger-than-average beak size exceed the benefits.

- At this point, finches with average beak size in the population will be the most likely to survive and reproduce, and finches with deeper or shallower beaks than the new average will be at a disadvantage.

At this point, the population reaches equilibrium with regard to beak size. The process that produces this equilibrium state is called stabilizing selection.

Even though average characteristics of the beak in the population will not change in this situation, selection is still going on. The point to remember here is that populations do not remain static over the long run; if so, it is because a population is consistently favored by stabilizing selection.

Rate of Evolutionary Change

In Darwin's day, the idea that natural selection could change a chimpanzee into a human, much less that it might do so in just a few million years (which is a brief moment in evolutionary time), was unthinkable.

Today, most scientists believe that humans evolved from an apelike creature in only 5 to 10 million years. In fact, some of the rates of selective change observed in contemporary populations are far faster than necessary for natural selection to produce the adaptations that we observe.

The human brain has roughly doubled in the last 2 million years (rate of change of 0.00005% per year); that is 10,000 times slower than the rate of change observed among finches in the Galapagos Islands.

Therefore the real puzzle is why the change in the fossil record seem to have been quite slow.

The fossil record is still very incomplete.

It is quite likely that some evolutionary changes in the past were rapid, but the sparseness of the fossil record prevents us from detecting them.

Darwin's Difficulties

In *On the Origin of Species*, Darwin proposed that new species and other major evolutionary changes arise by the accumulation of small variations through natural selection.

This idea was not widely embraced by his contemporaries.

- Many accepted the idea that new species arise through the transformation of existing species.

- Many accepted the idea that natural selection is the most important cause of organic change.

- But only a few endorsed Darwin's view that major changes occur through the accumulation of small variations.

Darwin's critics raised a major objection to his theory: The actions of selection would inevitably deplete variation in populations and make it impossible for natural selection to continue.

Yet Darwin couldn't convince his contemporaries that evolution occurred through the accumulation of small variations because he could not explain how variation is maintained, because he and his contemporaries did not yet understand the mechanics of inheritance.

For most people at the time, including Darwin, many of the characteristics of offspring were thought to be an average of the characteristics of their parents. This phenomena was believed to be caused by the action of blending inheritance, a model of inheritance that assumes the mother and father each contribute a hereditary substance that mixes, or "blends", to determine the characteristics of the offspring.

The solution to these problems required an understanding of genetics, which was not available for another half century. It was not until well into the 20th century that geneticists came to understand how variation is maintained, and Darwin's theory of evolution was generally accepted.

GENETICS: 19th AND 20th CENTURIES

Although Charles Darwin is credited with discovering the first observations of natural selection, he never explained how or why the process happens. Other scholars tackled these problems.

Gregor Mendel (1822-1884)

Darwin recognized the importance of individual variation in process of natural selection, but could not explain how individual differences were transmitted from one generation to another.

Although none of main scientists in the 19th-century debate about evolution knew it, the key experiments necessary to understand how genetic inheritance really worked had already been performed by an obscure monk, Gregor Mendel, who lived near Brno, in the Czech Republic.

Between 1856 and 1863, Mendel performed many breeding experiments using the common edible garden pea plants. He meticulously recorded his observations and isolated a number of traits in order to confirm his results.

In 1866, Mendel published a report where he described many features of the mode of inheritance which Darwin was seeking. He proposed the existence of three fundamental principles of inheritance: Segregation; Independent Assortment; Dominance and Recessiveness.

Because the basic rules of inheritance Mendel discovered apply to humans as well as to peas, his work is of prime relevance for paleoanthropology and human evolution.

Nevertheless Mendel's work was beyond the thinking of the time; its significance was overlooked and unrecognized until the beginning of the beginning of the 20th century.

Mendelian Genetics

Mendel's research

Mendel observed that his peas had seven easily observable characteristics, with only two forms, or variants, for each trait:

Seed texture	smooth	wrinkled
Seed interior color	yellow	green
Seed coat color	gray	white
Ripe pods	inflated	constricted
Unripe pods	green	yellow
Position of flowers on stem	along stem	end of stem

Length of stem	long	short

After crossing plants, Mendel noted and carefully recorded the number of plants in each generation with a given trait. He believed that the ratio of plant varieties in a generation of offspring would yield clues about inheritance, and he continually tested his ideas by performing more experiments.

From his controlled experiments and the large sample of numerous breeding experiments, Mendel proposed the existence of three fundamental principles of inheritance:

- Segregation

- Independent Assortment

- Dominance and Recessiveness

Segregation

Mendel began crossing different varieties of purebred plants that differed with regard to a specific trait. For example, pea color.

In the experiment:

- The first generation (parental, Fo) of plants were either green or yellow. As they matured, the first hybrid offspring generation was not intermediate in color, as blending theories of inheritance (Darwin) would have predicted. To the contrary, they were all yellow.

Next, Mendel allowed these plants to self fertilize and produce a second generation of plants (generation F1). But this time, only 3/4 of offspring plants were yellow, and the remaining 1/4 were green.

These results suggested an important fact:

Different expressions of a trait were controlled by discrete units, which occurred in pairs, and that offspring inherited one unit from each parent.

This is Mendel's first principle of inheritance: principle of segregation.

Independent Assortment

Mendel also made crosses in which two traits were considered simultaneously to determine whether there was a relationship between them. For example: Plant height and seed color.

Results of experiments: No relationship between the two traits were found; nothing to dictate that a tall plant must have yellow (or green) seeds; therefore, expression of one trait is not influenced by the expression of the other trait.

Based on these results, Mendel stated his second principle of inheritance: the principle of independent assortment. This principle says that the genes that code for different traits assort independently of each other.

Dominance and Recessiveness

Mendel also recognized that the trait that was absent in the first generation of offspring plants had not actually disappeared at all - it had remained, but was masked and could not be expressed.

To describe the trait that seemed to be lost, Mendel used the term recessive; the trait that was expressed was said to be dominant.

Thus the important principle of dominance and recessiveness was formulated; and it remains today an essential concept in the field of genetics.

Implications of Mendel's research

Mendel thought his findings were important, so he published them in 1866.

Scientists, especially botanists studying inheritance, in the late 19th century, should have understood the importance of Mendel's experiments. But instead, they dismissed Mendel's work, perhaps because it contradicted their own results or because he was an obscure monk.

Soon after the publication of his work, Mendel was elected abbot of his monastery and was forced to give up his experiments.

His ideas did not resurface until the turn of the 20th century, when several botanists independently replicated Mendel's experiments and rediscovered the laws of inheritance.

The role of cell division in inheretence

Mitosis and Meiosis

By the time Mendel's experiments were rediscovered in 1900, some facts were well known:

- virtually all living organisms are built out of cells;
- all the cells in complex organisms arise from a single cell through the process of cell division.

In order for plants and animals to grow and maintain good health, body cells of an organism must divide and produce new cells. Cell division is the process that results in the production of new cells.

Two types of cell division have been identified:

- **Mitosis**: a process when chromosomes (and genes) replicate, forming a second pair that duplicates the original pair of chromosomes in the nucleus. Thus, mitosis produces new cells (daughter cells) that have exactly the same number of chromosome pairs and genes, as did the parent cell

- **Meiosis**: while mitosis produces new cells (which contain a pair of homologous chromosomes), meiosis leads to development of new individuals, known as gametes (which contain only one copy of each chromosome). With this process, each new cell (containing only one copy of each chromosome) is said to be haploid: when new individual is conceived, a haploid sperm from father unites with a haploid egg from the mother to produce a diploid zygote. The zygote is a single cell that divides mitotically over and over again to produce the millions and millions of cells that make up an individual's body.

Mendel and chromosomes

Mendel stated in 1866 that an organism's observed traits are determined by "particules" (later named genes by the American geneticist T.H. Morgan) acquired from each of the parents. This statement was only understood by further research.

Between the time of Mendel's initial discovery of the nature of inheritance and its rediscovery at the turn of the century, a crucial feature of cellular anatomy was discovered: the chromosome.

Chromosomes are small, linear bodies contained in every cell and replicated during cell division.

In 1902, a graduate student from Columbia University, (Walter Sutton) made the connection between chromosomes and properties of inheritance discovered by Mendel's principles:

- genes reside on chromosomes because individuals inherit one copy of each chromosome from each parent

- therefore an organism's observed traits are determined by genes from both parents

- these propositions are consistent with the observation that mitosis transmits a copy of both chromosomes to every daughter cell, so every cell contains copies of both the maternal and paternal chromosomes.

Molecular genetics

In first half of the 20th century, geneticists made substantial progress in:

- describing the cellular events that took place during mitosis and meiosis

- understanding the chemistry of reproduction.

By the middle of the 20th century it was known that chromosomes contain two structurally complex molecules: protein and DNA (deoxyribonucleic acid). It was also determined that the particle of heredity postulated by Mendel was DNA, not protein - though exactly how DNA might contain and convey the information essential to life was still a mystery.

In the early 1950s, several biologists (led by Francis Crick and James Watson), at Cambridge University, made a discovery that revolutionized biology: they deduced the structure of DNA.

Through this discovery, we now know how DNA stores information and how this information controls the chemistry of life, and this knowledge explains why heredity leads to the patterns Mendel describes in pea plants, and why there are sometimes new variations.

Molecular Components

Cells

Cells are basic units of life in all living organisms. Complex multicellular forms (plants, insects, birds, humans, ...) are composed of billions of cells, all functioning in complex ways to promote the survival of the individual.

DNA   Molecules

Complex molecule with an unusual shape: like two strands (called nucleotides) of a rope (composed of alternating sequences of phosphate and sugar molecules) twisted around one another (double helix). Chemical bases that connect two strands constitute code that contains information to direct production of proteins.

It is at this level that development of certain traits occurs; . . Yet, since the DNA in a single chromosome is millions of bases long, there is room for a nearly infinite variety of messages.

DNA molecules have the unique property of being able to produce exact copies of themselves: as long as no errors are made in the replication process, new organisms will contain genetic material exactly like that in ancestral organisms.

Genes

A Gene is a short segment of the DNA molecule that directs the development of observable or identifiable traits. Thus genetics is the study of how traits are transmitted from one generation to the next.

Chromosomes

Each chromosome contains a single DNA molecule, roughly two meters long that is folded up to fit in the nucleus. Chromosomes are nothing more than long strands of DNA combined with protein to produce structures that can actually be seen under a conventional light microscope

Each kind of organism has characteristic number of chromosomes, which are usually found in pairs. For example, human cells contain 23 pairs.

Cellular processes

DNA   Replication

In addition to preserving a message faithfully, hereditary material must be replicable. Without the ability to make copies of itself, the genetic message that directs the activities of living cells could not be spread to offspring, and natural selection would be impossible.

Cells multiply by dividing in such a way that each new cell receives a full complement of genetic material. For new cells to receive the essential amount of DNA, it is first necessary for the DNA to replicate:

• Specific enzymes break the bonds between bases in the DNA molecule, leaving the two previously joined strands exposed

- When process is completed, there are two double-stranded DNA molecules exactly like the original one.

Protein   **Synthesis**
One of most important functions of DNA is that it directs protein synthesis within the cell. Proteins are complex, three-dimensional molecules that function through their ability to bind to other molecules.

Proteins function in myriad ways:

- **Collagen** is most common protein in body and major component of all connective tissues

- **Enzymes** are also proteins; their function is to initiate and enhance chemical reactions

- **Hormones** are another class.

Proteins are not only major constituents of all body tissues, but also direct and perform physiological and cellular functions. Therefore critical that protein synthesis occur accurately, for, if it does not, physiological development and metabolic activities can be disrupted or even prevented.

Evolutionary significance of cellular processes

Meiosis is a highly important evolutionary innovation, since it increases variation in populations at a faster rate than mutation alone can do in asexually reproducing species.

Individual members of sexually reproducing species are not genetically identical clones of other individuals. Therefore each individual represents a unique combination of genes that has never occurred before and will never occur again.

Genetic diversity is therefore considerably enhanced by meiosis. If all individuals in a population are genetically identical over time, the natural selection and evolution cannot occur. Therefore, sexual reproduction and meiosis are of major evolutionary importance because they contribute to the role of natural selection in populations.

Synthesizing the knowledge

Darwin believed that evolution proceeded by the gradual accumulation of small changes. But Mendel and the biologists who elucidated the structure of the genetic system around the turn of the century proved that inheritance was fundamentally discontinuous.

Yet turn-of-the-century geneticists argued that this fact could not be reconciled with Darwin's idea that adaptation occurs through the accumulation of small variations.

If generation of parent plants are tall and short, then there will be no intermediate in the generation of offspring and size of peas cannot change in small steps. In a population of short plants, tall ones must be created all at once by mutation, not gradually lengthened over time by selection.

These arguments convinced most biologists of the time, and consequently Darwinism was in decline during the early part of the 20th century.

In the early 1930s, a team of British and American biologists showed how Mendelian genetics could be used to explain continuous variation. Their insights led to the resolution of two main objections to Darwin's theory:

- the absence of a theory of inheritance
- the problem of accounting for how variation is maintained in populations

When their theory was combined with Darwin's theory of natural selection and with modern biological studies, a powerful explanation of organic evolution emerged. This body of theory and the supporting empirical evidence is now called the modern synthesis.

Variation maintained

Darwin knew nothing about genetics, and his theory of adaptation by natural selection was framed as a "struggle for existence": there is variation of observed traits that affects survival and reproduction, and this variation if heritable.

Also, the blending model of inheritance appealed to 19th century thinkers, because it explained the fact that for most continuously varying characters, offspring are intermediate between their parents.

When yellow and blue parents are crossed to produce a green offspring, the blending model posits that the hereditary material has mixed, so that when two green individuals mate they produce only green offspring.

According to Mendelian genetics, however, the effects of genes are blended in their expression to produce a green phenotype, but the genes themselves remain unchanged. Thus, when two green parents mate, they can produce blue, yellow and green offspring.

Sexual reproduction produces no blending in the genes themselves, despite the fact that offspring may appear to be intermediate between their parents. This is because genetic transmission involves faithful copying of the genes themselves and reassembling them in different combinations in zygotes.

The only blending that occurs takes place at the level of the expression of genes in phenotypes (ex. Beak depth, pea color). The genes themselves remain distinct physical entities.

Yet, these facts do not completely solve the problem of the maintenance of variation. Indeed, even is selection tends to deplete variation, there would still be variation of traits due to environmental effects. In fact, without genetic variation there can be no further adaptation.

Mutation

Genes are copied with amazing fidelity, and their messages are protected from random degradation by a number of molecular repair mechanisms.

However, every once in a while, a mistake in copying is made that goes unrepaired. These mistakes damage the DNA and alter the message that it carries.

These changes are called **mutations**, and they add variation to a population by continuously introducing new genes, some of which may produce novel traits that selection can assemble into adaptations. Although rates of mutation are very slow, this process plays an important role in generating variation.

More importantly, this process provides the solution to one of Darwin's dilemma: the problem of accounting for how variation is maintained in populations.

Twentieth century research has shown that there are two pools of genetic variation: hidden and expressed. Mutation adds new genetic variation, and selection removes it from the pool of expressed variation. Segregation and recombination shuffle variation back and forth between the two pools with each generation.

In other words: if individuals with a variety of genotypes are equally likely to survive and reproduce, a considerable amount of variation is protected (or hidden) from selection; and because of this process, a very low mutation rate can maintain variation despite the depleting action of selection.

Human evolution and adaptation are intimately linked to life processes that involve cells, replication and decoding of genetic information, and transmission of this information between generations. Because physical anthropologists are concerned with human evolution, adaptation, and variation, they must have a thorough understanding of the factors that lie at very root of these phenomena. Because it is genetics that ultimately links or influences many of the various subdisciplines of biological anthropology.

LIVING PRIMATES

The Classification System

In order to understand the exact place of humanity among the animals, it is helpful to describe system used by biologists to classify living things. The basic system was devised by 18th-century Swedish naturalist Carl von LinnÃ(c).

The purpose of the Linnean system was simply to create order in great mass of confusing biological data that had accumulated by that time. Von LinnÃ(c) classified living things on the basis of overall similarities into small groups or species. On the basis of homologies, groups of like species are organized into larger, more inclusive groups, called genera.

Through careful comparison and analysis, von LinnÃ(c) and those who have come after him have been able to classify specific animals into a series of larger and more inclusive groups up to the largest and most inclusive of all, the animal kingdom.

The Primate Order

Primates are only one of several mammalian orders, such as rodents, carnivores, and ungulates.

As such, primates share a number of features with other mammals:

- mammals are intelligent animals

- in most species, young are born live, egg being retained within the womb of the female until it achieves an advanced state of growth

- once born, the young are nourished by their mothers

- mammals have a constant body temperature, an efficient respiratory system featuring a separation between the nasal and mouth cavities, an efficient four-chambered heart that prevents mixing of oxygenated and deoxygenated blood, among other characterustics

- the skeleton of most mammals is simplified compared to that of most reptiles, in that it has fewer bones. For example, the lower jaw consists of a single bone, rather than several.

Species

In modern evolutionary biology, the term species is usually defined as a population or group of organisms that look more or less alike and that is potentially capable of interbreeding to produce fertile offspring. Practically speaking, individuals are usually assigned to a species based on their appearance, but it is their ability to interbreed that ultimately validates (or invalidates) the assignment. Thus, no matter how similar two populations may look, if they are incapable of interbreeding, they must be assigned to different species.

Populations within a species that are quite capable of interbreeding but may not regularly do so are called races or subspecies. Evolutionary theory suggests that species evolve from races through the accumulation of differences in the gene pools of the separated groups.

Primate Characteristics

Although living primates are a varied group of animals, they do share a number of features in common. These features are displayed in varying degrees by the different kinds of primates: in some they are barely detectable, while in others they are greatly elaborated.

All are useful in one way or another to arboreal (or tree-dwelling) animals, although they are not essential to life in trees.

Primate Sense Organs

The primates' adaptation to their way of life in the trees coincided with changes in the form and function of their sensory apparatus: the senses of sight and touch became highly developed, and the sense of smell declined.

Catching insects in trees, as the early primates did and as many still do, demands quickness of movement and the ability to land in the right place without falling. Thus, they had to be adept at judging depth, direction, distance and the relationship of objects in space.

Primate sense of touch became also highly developed as a result of arboreal living. Primates found useful an effective feeling and grasping mechanism to grab their insect prey, and to prevent them from falling and tumbling while moving through the trees.

The Primate Brain

By far the most outstanding characteristic of primate evolution has been the enlargement of the brain among members of the order. Primate brains tend to be large, heavy in proportion to body weight, and very complex.

Reasons for this important change in brain size are many:

- Prior to 65 Myrs ago, mammals seem to have been nocturnal in their habits; after 65 million years ago, primates began to carry out their activities in the daylight hours. As a consequence, the sense of vision took on greater importance, and so visual acuity was favored by natural selection

- another hypothesis involves the use of the hand as a tactile organ to replace the teeth and jaws. The hands assumed some of the grasping, tearing and dividing functions of the snout, again requiring development of the brain centers for more complete coordination.

- The enlarged cortex not only provided the primates with a greater degree of efficiency in the daily struggle for survival but also gave them the basis for more sophisticated cerebration or thought. The ability to think probably played a decisive role in the evolution of the primates from which human beings emerged.

Primate Teeth

Although they have added other things than insects to their diets, primates have retained less specialized teeth than other mammals.

The evolutionary trend for primate dentition has generally been toward economy, with fewer, smaller, more efficient teeth doing more work.

Our own 32 teeth are fewer in number than those of some, and more generalized than most, primates.

Primate Skeleton

A number of factors are responsible for the shape of the primate skull as compared with those of most other mammals: changes in dentition, changes in the sensory organs of sight and smell, and increase in brain size. As a result, primates have more a humanlike face than other mammals.

The upper body is shaped in such ways to allow greater maneuverability of the arms, permitting them to swing sideways and outward from the trunk of the body.

The structural characteristics of the primate foot and hand make grasping possible; the digits are extremely flexible, the big toe is fully opposable to the other digits in most species, and the thumb is opposable to the other digits to varying degrees. The flexible, unspecialized primate hand was to prove a valuable asset for future evolution of this group. It allowed early hominines to manufacture and utilize tools and thus embark on the new and unique evolutionary pathway that led to the revolutionary ability to adapt through culture.

Types of Living Primates

Prosimians

The most primitive of the primates are represented by the various prosimians, including the lemurs and the lorises, which are more similar anatomically to earlier mammalian ancestors than are other primates (monkeys, apes, humans). They tend to exhibit certain more ancestral features, such as a more pronounced reliance on olfaction (sense of smell). Their greater olfactory capabilities are reflected in presence of moist, fleshy pad at end of nose and in relatively long snout.

Lemurs and lorises represent the same general adaptive level. Both groups exhibit good grasping and climbing abilities and a fairly well developed visual apparatus, although vision is not completely stereoscopic, and color vision may not be as well developed as in anthropoids.

Lemurs

At present, lemurs are found only on the island of Madagascar and adjacent islands off the east coast of Africa.

As the only natural nonhuman primates on this island, they diversified into numerous and varied ecological niches without competition from monkeys and apes. Thus, the 52 surviving species of

Madagascar represent an evolutionary pattern that has vanished elsewhere.

Lemurs range in size from 5 inches to a little over two feet. While the larger lemurs are diurnal and exploit a wide variety of dietary items (leaves, fruits, buds, bark), the smaller forms (mouse and dwarf lemurs) are nocturnal and insectivorous.

Lemurs display considerable variation regarding numerous other aspects of behavior. While many are primarily arboreal, others (e.g., ring-tailed lemur) are more terrestrial. Some arboreal species are quadrupeds, and others are vertical clingers and leapers.

Lorises

Lorises are similar in appearance to lemurs, but were able to survive in mainland areas by adopting a nocturnal activity pattern at a time when most other prosimians became extinct. Thus, they were (and are still) able to avoid competition with more recently evolved primates (dirunal monkeys).

There are five loris species, all of which are found in tropical forest and woodland habitats of India, Sri Lanka, Southeast Asia and Africa.

Locomotion in lorises is a slow, cautious climbing form of quadrupedalism, and flexible hip joints permit suspension by hind limbs while the hands are used in feeding. Some lorises are almost entirely insectivorous; others supplement their diet with various combinations of fruits, leaves, gums, etc.

Tarsiers

There are seven recognized species, all restricted to island areas in Southeast Asia. They inhabit a wide range of forest types, from tropical forest to backyard gardens.

They are nocturnal insectivores, leaping onto prey from lower branches and shrubs. They appear to form stable pair bonds, and the basic tarsier social unit is a mated pair and their young offspring.

Tarsiers present complex blend of characteristics not seen in other primates. They are unique in that their enormous eyes, which dominate much of the face, are immobile within their sockets. To compensate for this inability to move the eyes, tarsiers are able to rotate their heads 180Â° in an owl-like manner.

Simians

Although there is much variation among simians (also called anthropoids), there are certain features that, when taken together, distinguish them as a group from prosimians (and other mammals)

- generally larger body size
- larger brain

- reduced reliance on the sense of smell

- increased reliance on vision, with forward-facing eyes placed at front of the face

- greater degree of color vision

- back of eye socket formed by a bony plate

- blood supply to brain different from that of prosimians

- fusion of two sides of mandible at midline to form one bone

- less specialized dentition

- differences with regard to female internal reproductive anatomy

- longer gestation and maturation periods

- increased parental care

- more mutual grooming

Monkeys

Approximately 70 percent of all primates (about 240 species) are monkeys, although it is frequently impossible to give precise numbers of species because taxonomic status of some primates remains in doubt and constant new discoveries.

Monkeys are divided into two groups (New World and Old World) separated by geographical area as well as by several million years of separate evolutionary history.

New World monkeys exhibit wide range of size, diet, and ecological adaptation. In size, they vary from tiny marmosets and tamarins to the 20-pound howler monkey. Almost all are exclusively arboreal; most are dirunal. Although confined to trees, New World monkeys can be found in wide range of arboreal environments throughout most forested areas in Southern Mexico and Central and South America. One of characteristics distinguishing New World monkeys from Old World is shape of nose: they have broad noses with outward-facing nostrils.

Old World monkeys display much more morphological and behavioral diversity than New World monkeys. Except for humans, they are the most widely distributed of all living primates. They are found throughout sub-Saharan Africa and Southern Asia, ranging from tropical jungle habitats to semiarid desert and even to seasonally snow-covered areas in northern Japan. Most are quadrupedal and primarily arboreal.

Apes and humans

This group is made of several families:

- Hylobatidae (gibbons and siamangs)

- Pongidae (orangutans)

- Hominidae (humans, gorillas, common chimapanzees, bonobos)

They differ from monkeys in numerous ways:

- generally larger body size, except for gibbons and siamangs

- absence of a tail

- shortened trunk

- differences in position and musculature of the shoulder joint (adapted for suspensory locomotion)

- more complex behavior;

- more complex brain and enhanced cognitive abilities;

- increased period of infant development and dependency.

Orangutans

Found today only in heavily forested areas on the Indonesian islands of Borneo and Sumatra, orangutans are slow, cautious climbers whose locomotor behavior can best be described as "four-handed", a tendency to use all four limbs for grasping and support. Although they are almost completely arboreal, they do sometimes travel quadrupedally on ground. They are very large animals with pronounced sexual dimorphism: males weigh over 200 pounds while females are usually less than 100 pounds.

Gorillas

The largest of all living primates, gorillas are today confined to forested areas of western and equatorial Africa. There are four generally recognized subspecies: Western Lowland Gorilla, Cross River Gorilla, Eastern Lowland Gorilla, and Mountain Gorilla. Gorillas exhibit strong sexual dimorphism. Because of their weight, adult gorillas, especially males, are primarily terrestrial and adopt a semiquadrupedal (knuckle-walking) posture on the ground. All gorillas are almost exclusively vegetarian.

Common Chimpanzees

The best-known of all nonhuman primates, Common Chimpanzees are found in equatorial Africa. In many ways, they are structurally similar to gorillas, with corresponding limb proportions and upper body shape: similarity due to commonalities in locomotion when on the ground (quadrupedal knuckle-walking). However, chimps spend more time in trees; when on the ground, they frequently walk bipedally for short distances when carrying food or other objects.

They are highly excitable, active and noisy. Common Chimpanzee social behavior is complex, and individuals form lifelong attachments with friends and relatives. They live in large, fluid communities of as many as 50 individuals or more. At the core of a community is a group of bonded males. They act as a group to defend their territory and highly intolerant of unfamiliar chimps, especially nongroup males.

Bonobos

Found only in an area south of the Zaire River in the Democratic Republic of Congo, Bonobos (also called Pygmy Chimpanzees) exhibit strong resemblance to Common Chimpanzees, but are somewhat smaller. Yet they exhibit several anatomical and behavioral differences. Physically, they have a more linear body build, longer legs relative to arms, relatively smaller head, dark face from birth. Bonobos are more arboreal than Common Chimpanzees, and they appear to be less excitable and aggressive.

Like Common Chimpanzees, Bonobos live in geographically based, fluid communities, and they exploit many of the same foods, including occasional meat derived from killing small mammals. But they are not centered around a group of closely bonded males. Instead, male-female bonding is more important than in Common Chimpanzees.

WHAT MAKES A PRIMATE HUMAN?

1. *What are the implications of the shared characteristics between humans and the other primates?*

2. *Why do anthropologists study the social behavior of monkeys and apes?*

Information about primate behavior and ecology plays an integral role in the story of human evolution.

1. Humans are primates, and the first members of the human species were probably more similar to living nonhuman primates than to any other animals on earth. Thus, by studying living primates we can learn something about the lives of our ancestors.

2. Humans are closely related to primates and similar to them in many ways. If we understand how evolution has shaped the behavior of animals so much like ourselves, we may have greater insights about the way evolution has shaped our own behavior and the behavior of our ancestors.

Primate social behavior

Over the past four decades, primatologists have made prolonged close-range observations of monkeys and apes in their natural habitats, and we are discovering much about social organization, learning ability, and communication among our closest relatives (chimpanzees, and gorillas) in the animal kingdom.

In particular, we are finding that a number of behavioral traits that we used to think of as distinctively human are found to one degree or another among other primates, reminding us that many of the differences between us and them are differences of degree, rather than kind.

The Group

Primates are social animals, living and traveling in groups that vary in size from species to species. In most species, females and their offspring constitute core of social system.

Among chimps, the largest organizational unit is the community, composed of 50 or more individuals. Rarely however are all these animals together at a single time. Instead they are usually ranging singly or in small subgroups consisting of adult males together, females with their young, or males and females together with their young. In the course of their travels, subgroups may join forces and forage together, but sooner or later these will break up into smaller units.

Dominance

Many primate societies are organized into dominance hierarchies, that impose some degree of order with groups by establishing parameters of individual behavior.

Although aggression is frequently a means of increasing one's status, dominance usually serves to reduce the amount of actual physical violence. Not only are lower-ranking animals unlikely to attack or even threaten a higher-ranking one, but dominant animals are also frequently able to exert control simply by making a threatening gesture.

Individual rank or status may be measured by access to resources, including food items and mating partners.

An individual's rank is not permanent and changes throughout life. It is influenced by many factors, including sex, age, level of aggression, amount of time spent in the group, intelligence, etc.

In species organized into groups containing a number of females associated with one or several adult males, the males are generally dominant to females. Within such groups, males and females have separate hierarchies, although very high ranking females can dominate the lowest-ranking males (particularly young ones).

Yet many exceptions to this pattern of male dominance:

- Among many lemur species, females are the dominant sex
- Among species that form monogamous pairs (e.g., indris, gibbons), males

and females are codominant

Aggression

Within primate societies, there is an interplay between affiliative behaviors that promote group cohesion and aggressive behaviors that can lead to group disruption. Conflict within a group frequently develops out of competition for resources, including mating partners and food items. Instead of actual attacks or fighting, most intragroup aggression occurs in the form of various signals and displays, frequently within the context of dominance hierarchy. Majority of such situations are resolved through various submissive and appeasement behaviors.

But conflict is not always resolved peacefully.

- High-ranking female macaques frequently intimidate, harass, and even attack lower-ranking females, particularly to restrict their access to food
- Competition between males for mates frequently results in injury and occasionally in death
- Aggressive encounters occur between groups as well as within groups
- Aggression occurs in the defense of territories

Individual interaction

To minimize actual violence and to defuse potentially dangerous situations, there is an array of affiliative, or friendly, behaviors that serve to reinforce bonds between individuals and enhance group stability. Common affiliative behaviors include reconciliation, consolation, and simple interactions between friends and relatives.

Most such behaviors involve various forms of physical contact including touching, hand holding, hugging, and, among chimpanzees, kissing. In fact, physical contact is one of the most important factors in primate development and is crucial in promoting peaceful relationships in many primate social groups.

One of the most notable primate activities is grooming, the ritual cleaning of another animal's coat to remove parasites, shreds of grass or other matter. Among bonobos and chimps, grooming is a gesture of friendliness, submission, appeasement or closeness.

The mother-infant bond is the strongest and most long-lasting in the group. It may last for many years; commonly for the lifetime of the mother.

Play

Frequent play activity among primate infants and juveniles is a means of learning about the environment, testing strength, and generally learning how to behave as adults. For example, Chimpanzee infants mimic the food-getting activities of their mothers, "attack" dozing adults, and "harass" adolescents.

Communication

Primates, like many animals, vocalize. They have a great range of calls that are often used together with movements of the face or body to convey a message.

Observers have not yet established the meaning of all the sounds, but a good number have been distinguished, such as warning calls, threat calls, defense calls, and gathering calls. Much of the communication takes place by the use of specific gestures and postures.

Home range

Primates usually move about within circumscribed areas, or home ranges, which are of varying sizes, depending on the size of the group and on ecological factors, such as availability of food. Home ranges are often moved seasonally. The distance traveled by a group in a day varies, but may include many miles.

Within this home range is a portion known as the core area, which contains the highest concentration of predictable resources (water, food) and where the group is most frequently found (with resting places and sleeping trees).

The core area can also be said to be a group's territory, and it is this portion of the home range that is usually defended against intrusion by others:

- Gorillas do not defend their home ranges against incursions of others of their kind

- Chimps, by contrast, have been observed patrolling their territories to ward off potential

101

trespassers

Among primates in general, the clearest territoriality appears in forest species, rather than in those that are terrestrial in their habits.

Tool use

A tool may be defined as an object used to facilitate some task or activity. A distinction must be made between simple tool use and tool making, which involves deliberate modification of some material for its intended use.

In the wild, gorillas do not make or use tools in any significant way, but chimpanzees do. Chimps modify objects to make them suitable for particular purposes. They can also pick up and even prepare objects for future use at some other location, and they can use objects as tools to solve new and novel problems.

Examples:

- use of stalks of grass to collect termites

- use of leaves as wipes or sponges to get water out of a hollow to drink

- use of rocks as hammers and anvils to open palm nuts and hard fruits

Primates and human evolution

Studies of monkeys and apes living today [especially those most closely related to humans: gorillas, bonobos and chimpanzees] provide essential clues in the reconstruction of adaptations and behavior patterns involved in the emergence of our earliest ancestors.

These practices have several implications:

- Chimpanzees can be engaged in activities that prepare them for a future (not immediate) task at a somewhat distant location. These actions imply planning and forethought

- Attention to the shape and size of the raw material indicates that chimpanzee toolmakers have a preconceived idea of what the finished product needs to be in order to be useful

To produce a tool, even a simple tool, based on a concept is an extremely complex behavior. Scientists previously believed that such behavior was the exclusive domain of humans, but now we must question this very basic assumption.

At the same time, we must be careful about how we reconstruct this development. Primates have changed in various ways from earlier times, and undoubtedly certain forms of behavior that they now exhibit were not found among their ancestors.

Also it is important to remember that present-day primate behavior shows considerable variation, not just from one species to another, but also from one population to another within a single species.

Primate fossils

The study of early primate fossils tells us something we can use to interpret the evolution of the entire primate line, including ourselves. It gives us a better understanding of the physical forces that caused these primitive creatures to evolve into today's primates.

Ultimately, the study of these ancient ancestors gives us a fuller knowledge of the processes through which insect-eating, small-brained animals evolved into a toolmaker and thinker that is recognizably human.

Rise of the primates

For animals that have often lived where conditions for fossilization are generally poor, we do have a surprisingly large number of primate fossils. Some are relatively complete skeletons, while most are teeth and jaw fragments.

Primates arose as part of a great adaptive radiation that began more than 100 million years after the appearance of the first mammals. The reason for this late diversification of mammals was that most ecological niches that they have since occupied were either preempted by reptiles or were nonexistent until the flowering plants became widespread beginning about 65 million years ago.

By 65 million years ago, primates were probably beginning to diverge from other mammalian lineages (such as those which later led to rodents, bats and carnivores). For the period between 65-55 Myrs ago (Paleocene), it is extremely difficult to identify the earliest members of the primate order:

- available fossil material is scarce

- first primates were not easily distinguished from other early (generalized) mammals

Eocene primates

First fossil forms that are clearly identifiable as primates appeared during Eocene (55-34 million years ago). From this period have been recovered a wide variety of primates, which can all be called prosimians. Lemur-like adapids were common in the Eocene, as were species of tarsier-like primates.

These first primates were insect eaters and their characteristics developed as an adaptation to the initial tree-dwelling environment:

- larger, rounder braincases

- nails instead of claws

- eyes that are rotated forward, allowing overlapping fields of perception and thus binocular vision

- presence of opposable large toe

This time period exhibited the widest geographical distribution and broadest adaptive radiation ever displayed by prosimians. In recent years, numerous finds of Late Eocene (36-34 Myrs ago) suggest that members of the adapid family were the most likely candidates as ancestors of early anthropoids.

Oligocene primates

The center of action for primate evolution after Eocene is confined largely to Old World. Only on the continents of Africa and Eurasia can we trace the evolutionary development of apes and hominids due to crucial geological events; particularly continental drift.

During the Oligocene (34-23 Myrs ago), great deal of diversification among primates occurred. The vast majority of Old World primate fossils for this period comes from just one locality: the Fayum area of Egypt. From Fayum, 21 different species have been identified.

Apidium

- Most abundant of all Oligocene forms

- This animal was about size of squirrel

- Teeth suggest diet composed of fruits and probably seeds

- Preserved remains of the limbs indicate that this creature was a small arboreal quadruped, adept at leaping and springing

Propliopithecus

- Morphologically, quite primitive

- Not showing particular derived tendencies in any direction

- Small to medium in size

- Likely fruit eaters

Aegyptopithecus

- Largest of Fayum anthropoids, roughly the size of modern howler (13 to 18 pounds)

- Primitive skull, which is small and resembles the one of a monkey in some details

- From analysis of limb proportions and muscle insertions, Aegyptopithecus was a short-limbed, heavily muscled, slow-moving arboreal quadruped

- This form of primates important because, better than any other fossil primate, it bridges the gap between the Eocene prosimians and the Miocene hominoids.

A great abundance of hominoid fossil material has been found in the Old World from the Miocene (23-7 million years ago).

Based on size, these fossils can be divided into two major subgroupings: small-bodied and large-bodied hominoids.

- Small-bodied varieties comprise gibbon and siamang

- Large-bodied hominoids are Pongo (orangutan), Gorilla, Pan (chimpanzees and bonobos) and Homo. These four forms can then be subdivided into two major subgroups: Asian large-bodied (orangutan) and African large-bodied (gorillas, chimpanzees, bonobos, and humans).

The remarkable evolutionary success represented by the adaptive radiation of large-bodied hominoids is shown in its geographical range from Africa to Eurasia. Large-bodied hominoids first evolved in Africa around 23 million years ago. Then they migrated into Eurasia, dispersed rapidly and diversified into a variety of species. After 14 million years ago, we have evidence of widely distributed hominoids in many parts of Asia and Europe. The separation of the Asian large-bodied hominoid line from the African stock (leading ultimately to gorillas, chimps and humans) thus would have occurred at about that time.

African Forms

- A wealth of early hominoid fossils has come from deep and rich stratigraphic layers of Kenya and Uganda

- These diverse forms are presently classified into at least 23 species of hominoids

- Best samples and thus best known forms are those of genus Proconsul

- Environmental niches were quite varied: some species were apparently confined to dense rain forests, others potentially exploited more open woodlands

- Considerable diversity of locomotor behaviors: some were at least partially terrestrial - on the ground, some may have even occasionally adopted a bipedal position

- Most forms were probably fruit eaters, some may also have included considerable amounts of leaves in their diet

European Forms

- Very few fossils have been discovered

- Most researchers would place these forms into the genus Dryopithecus

- Evolutionary relationship with other hominoids is both difficult and controversial at this point.

South/Southwest Asian Forms

- Remains have been found in Turkey, India and Pakistan

- Attributed to genus Sivapithecus

- Probably good-sized hominoid (70-150 pounds), that inhabited a mostly arboreal niche

- Facial remains have concave profiles and projecting incisors, which bear striking similarities with the orangutan

- Other traits are distinctively unlike an orangutan. For example, the forelimb suggests a unique mixture of traits, indicating probably some mode of arboreal quadrupedalism but with no suspensory component.

Miocene apes and Human Origin

The large-bodied African hominoids appeared by 16 million years ago and were widespread even as recently as 8 million years ago.

Based on fossils of teeth and jaws, it was easy to postulate some sort of relationship between them and humans. A number of features: position of incisors, reduced canines, thick enamel of the molars, and shape of tooth row, seemed to point in a somewhat human direction.

Although the African hominoids display a number of features from which hominine characteristics may be derived, and some may occasionally have walked bipedally, they were much too apelike to be considered hominines.

Nevertheless, existing evidence allows the hypothesis that apes and humans separated from a common evolutionary line sometime during the Late Miocene and some fossils, particularly the African hominoids, do possess traits associated with humans.

Not all African apes evolved into hominines. Those that remained in the forests and woodlands continued to develop as arboreal apes, although ultimately some of them took up a more terrestrial life. These are the bonobos, chimpanzees and gorillas, who have changed far more from the ncestral condition than have the still arboreal orangutans.

Chapter Summary

When did the first primates appear, and what were they like?

The earliest primates had developed by 60 million years ago and were small, arboreal insect eaters. Their initial adaptation to life in trees set the stage for the subsequent appearance of other primate models.

When did the first monkeys and apes appear, and what were they like?

By the Late Eocene (about 37 Myrs ago), monkeys and apes about the size of modern house cats were living in Africa. By about 20 million years ago, they had proliferated and soon spread over many parts of the Old World. Some forms remained relatively small, while others became quite large, some even larger than present-day gorillas.

When did group of primates give rise to the human line of evolution?

Present evidence suggests that our own ancestors are to be found among the African large-bodied hominoids, which were widespread between approximately 17 and 8 million years ago. Some of these ape-like primates lived in situations in which the right kind of selective pressure existed to transform them into primitive hominines.

MODERN HUMAN BEHAVIOR: ORIGIN OF LANGUAGE

Recognition of symbol use in the archaeological record (following Philip Chase's criteria):

- Regularity of use indicating purposeful and repeated activity;

- Yet repetitive behavior alone is not enough, because by itself not indicative of symbol use: Ex. Actions of individuals working without a system of shared meaning

- Therefore patterns need to be complex and learned linguistically rather than by observation or mimicry;

- Finally such behavior becomes symbolic, if it intentionally communicates thoughts, emotions, belief systems, group identity, etc. Material expressions of culturally mediated symbols:

- intentional burial of the dead, with grave goods;

- figurative and abstract imagery;

- pigment use;

- body ornamentation.

Important effort to clarify definition and category of data we are dealing with. Yet if we follow these criteria, most artefacts from Lower (Acheulean) and Middle Paleolithic (before 60,000-50,000 BP) are ruled out, because of lack of evidence for repeated patterning and intentionality.

Contribution of evolutionary psychology to origins of art

Is intelligence a single, general-purpose domain or a set of domains? Evolutionary psychologists answer: set of domains, which they call "mental modules", "multiple intelligences", "cognitive domains"; these "mental modules" interact, are connected;

Anatomically modern humans have better interaction between modules that other animals; therefore, able to perform more complex behaviors;

Four cognitive and physical processes exist:

- making visual images

- classification of images into classes

- intentional communication

- attribution of meaning to images

The first three are found in non-human primates and most hominids. Yet, only modern humans seem to have developed the fourth one.

For Neanderthals, intentional communication and classification were probably sealed in social intelligence module, while mark-making and attribution of meaning (both implicating material objects) were hidden. Only with arrival of modern humans, connection between modules made art possible by allowing intentional communication to escape into the domain of mark-making.

Problems with data and chronology

We could easily look at this transition in a smooth way: The passage from one industry to the next, one hominid to the next, etc. Evolutionary paths well structured and detailed, as in textbooks, but a bit too clear-cut, that is simplistic and reductionist.

After 1.8 million years ago, when *H. ergaster/erectus* moved out-of-Africa, the picture of human evolution becomes much more complex.

Situation due to several reasons:

- many more hominid species appear connected to global colonization and relative isolation;

- many cultural variations observed, illustrated by various stone tool industries, subsistence patterns, etc.

Overall, presence of differentiated cultural provinces in Africa and Eurasia which have their own evolutionary pace.

Dates don't seem to reveal a clear-cut divide between the Lower and Middle Paleolithic and don't fit anymore in a specific and rigorous time frame.

- H. erectus disappeared in most places around 300,000-200,000 yrs ago, although still found in Java up to 50,000 yrs ago;

- Archaic modern humans (Neanderthals) appeared around 130,000 yrs ago in Europe;

- Archaic modern humans (H. sapiens sapiens) appeared some time between 200,000 and 100,000 yrs ago in Africa;

- Acheulean stone tools were still in use beyond 200,000 yrs ago in many areas;

- The lithic industry (Mousterian) characteristic of the Middle Paleolithic appeared around 250,000 yrs ago in some areas (SW Asia);

- Subsistence patterns (hunting/scavenging), use of fire, habitats were still the basis of cultural adaptations in the Middle Paleolithic.

By focusing on a transition happening only at 50,000 yrs ago would be overlooking some major human innovations and evolutionary trends that took place earlier and on a much longer period.

We need to focus more on *H. heidelbergensis* and its material culture and other behavioral patterns to realize that the transition was not at 50,000 years ago, but between 600,000 and 60,000 yrs ago.

The revolution that wasn't

"Revolution" is in this context the Upper Paleolithic Revolution, with the development after 50,000 BCE of *Homo sapiens sapiens*, considered the only species anatomically AND behaviorally modern.

By "modern human behavior," we mean:

- Increased artifact diversity;

- Standardization of artefact types;

- Blade Technology;

- Worked bone and other organic materials;

- Personal ornaments and "art" or images;

- Structured living spaces;

- Ritual;

- Economic intensification, reflected in the exploitation of aquatic or other resources that require specialized technology;

- Expanded exchange networks.

By overlooking and even not considering recent discoveries from the 1990s regarding the periods before 50,000 years ago, we are misled to consider the evidence after that date as the result of biological and cultural revolution.

Recent observations in Africa, Europe and Asia from sites between 600,000 and 250,000 years ago (Acheulean period) seem to document very different patterns: "The Revolution That Wasn't".

Evidence in the Lower Paleolithic

STONE TOOLS: BLADE TECHNOLOGY

- Blade technology appeared and disappeared at different points in time. Earliest evidence to date: Kapthurin Formation (Kenya) 550,000-300,000 BP

BONE TOOLS

- Swartkrans (South Africa)

- Makapansgat (South Africa)
- Drimolen (South Africa)

WOODEN TOOLS

- Schöningen (Germany) 400,000 BP: Spears

USE OF PIGMENTS (OCHRE)

- Kapthurin Formation (Kenya) 550,000-300,000 BCE
- Twin Rivers Cave (Zambia) 400,000-200,000 BCE
- Pomongwe (Zimbabwe) 250,000-220,000 BCE
- Terra Amata (France) 300,000 BCE
- Becov (Czech Republic) 250,000 BCE

ARTISTIC EXPRESSION

- Pech de l'Azé (France) 400,000 BP: Engraved bone
- Sainte-Anne I Cave (France): Engraved bone
- Bilzingsleben (Germany) 300,000 BP: Large engraved rib
- Singi Talav (India) 300,000-150,000 BP: Occurrence of non-utilitarian objects (Quartz crystals)
- Zhoukoudian (China): Occurrence of non-utilitarian objects
- Birket Ram (Israel): Human figurine
- Olduvai Gorge (Tanzania): Figurine
- Makapansgat (South Africa): Human figurine
- Tan-Tan (Morocco) 500,000-300,000 BP: Human figurine

MORTUARY PRACTICES

- Atapuerca (Spain) 350,000 BP: H. heidelbergensis

SEAFARING

- Flores Island (Indonesia) 780,000 BP

Origins of language

Sometime during the last several million years, hominids evolved the ability to communicate much more complex and detailed information (about nature, technology, and social relationships) than any other creatures.

Yet we cannot reconstruct the evolutionary history of language as we reconstruct the history of bipedalism because the ability to use language leaves no clear traces in the fossil record. Therefore, there is no consensus among paleoanthropologists about when language evolved.

We are going to try to clarify the current situation by reviewing the recent evidence on the topic, focusing on specific criteria that could reveal essential information on early forms of language:

- brain capacity

- brain asymmetry

- vocal apparatus

The intellectual and linguistic skills of early hominids

Australopithecines

Reconstruction work on australopithecines indicates that their vocal tract was basically like that of apes, with the larynx and pharynx high up in the throat. This would not have allowed for the precise manipulation of air that is required for modern human languages. The early hominids could make sounds, but they would have been more like those of chimpanzees.

H. ergaster/erectus

Brain capacity

Their average cranial capacity was just a little short of the modern human minimum, and some individual erectus remains fall within the human modern range. It is difficult to be certain what this fact means in terms of intelligence.

Brain asymmetry

Paleoanthropologist Ralph Holloway has looked at the structure of *H. erectus* brains. He made endocasts of the inside surfaces of fossil crania, because the inside of the skull reflects some of the features of the brain it once held.

One intriguing find is that the brains of *H. erectus* were asymmetrical: the right and left halves of the brain did not have the same shape. This is found to a greater extent in modern humans, because the halves of our brains perform different functions. Language and the ability to use symbols, for example, are functions of our left hemispheres, while spatial reasoning (like the hand-eye coordination needed to make complex tools) is performed by the right hemisphere. This hints that *H. erectus* also had hemisphere specialization, perhaps even including the ability to communicate through a symbolic language.

Further evidence of language use by *H. erectus* is suggested by the reconstruction of the vocal apparatus based on the anatomy of the cranial base. Even though the vocal apparatus is made up of soft parts, those parts are connected to bone; so the shape of the bone is correlated with the shape of the larynx, pharynx and other features. *H. erectus* had vocal tracts more like those of modern humans, positioned lower in the throat and allowing for a greater range and speed of sound production. Thus, erectus could have produced vocal communication that involved many sounds with precise differences.

Whether or not they did so is another question. But given their ability to manufacture fairly complex tools and to survive in different and changing environmental circumstances, *H. ergaster/erectus* certainly could have had complex things to "talk about". Therefore it is not out of question that erectus had a communication system that was itself complex, even though some scholars are against this idea.

Summary

Scientists struggle with the definition of human behavior, while dealing with evidence dating to the early part of the Lower Paleolithic (7-2 million years ago).

Definition of modern human behavior is not easier to draw. The answer to this topic should not be found only in the period starting at around 50,000 yrs ago. Evidence now shows that the period between 500,000 and 250,000 years ago was rich in attempts at elaborating new behavioral patterns, either material or more symbolic.

On another level, beginning about 1.6 million years ago, brain size began to increase over and beyond that which can be explained by an increase in body size. Some researchers point to evidence that suggests that from 1.6 million years to about 300,000 years ago, the brain not only dramatically increased in size but also was being neurally reorganized in a way that increased its ability to process information in abstract (symbolic) way. This symbolism allowed complex information to be stored, relationships to be derived, and information to be efficiently retrieved and communicated to others in various ways.

Before 200,000 yrs ago, what is the relationship between *H. erectus* and *H. heidelbergensis*?

H. heidelbergensis seems to be the author of these new behavioral patterns, not *H. erectus*. *H. heidelbergensis*, especially in Africa, shows therefore evidence of new stone tool technology (blades), grinding stone and pigment (ochre) processing before 200,000 years ago. These new patterns connected with *H. heidelbergensis* could therefore be seen as critical advantages over *H. erectus* in the human evolutionary lineage.

References

- *How Humans Evolved*, Robert Boyd and Joan B. Silk, (1997)

- *Biological Anthropology*, Michael Park, (2002)

- *Physical Anthropology*, Philip L. Stein and Bruce M Rowe, (2003)

FROM HUNTER-GATHERERS TO FOOD PRODUCERS

Food Production

The ways in which humans procure resources are not unlimited. Essentially, there are five major procurement patterns practiced in the world today:

- Food collection
- hunting and gathering
- Food production
- extensive agriculture
- intensive agriculture
- pastoralism
- industrialism

Food Collection: Hunting and Gathering

People who practice a hunting and gathering subsistence strategy simply rely on whatever food is available in their local habitat, for the most part collecting various plant foods, hunting wild game, and fishing (where the environment permits).

They collect but they do not produce any food. For example, crops are not cultivated and animals are not kept for meat or milk Today, only about 30,000 people make their living in this fashion.

Cultures of <u>agriculturalists</u>, having larger <u>ecological footprints</u> have pushed most hunters and gatherers out of the areas where plant food and game is abundant into the more marginal of the earth: the Arctic, arid deserts, and dense tropical rain forests.

Food Production: Terminology

FOOD PRODUCTION: General term which covers types of domestication involving both plants and animals, each of which requires radically different practices.

CULTIVATION: Term refers to all types of plant culture, from slash-and-burn to growing crop trees. Terminological distinctions within the term cultivation are based on types on gardens maintained and means with which they are cultivated. Example: distinction between horticulture and agriculture

Horticulture: Refers to smaller-scale, garden-based cultivation, usually of a mixed variety of plant species, often with relatively simple tools.

Agriculture: This practice requires tools of greater complexity or higher energy in their manufacture and use, such as animal traction, etc.

Slash-and-burn: Strategy, normally horticultural, in which forest or bush land is cleared by chopping and burning the less useful wood species, planting in the ashes, harvesting for several years and then moving on to a new plot of land.

"Non-domestication" vs. "pre-domestication" cultivation: Cultivation of crops in some cases does not induce domestication. Example of such methods common among hunter-gatherers: beating the plants or reaping them when they are ripe. Therefore called "non-domestication cultivation." Other methods can induce the domestication of wild-type crops: uprooting or reaping grasses not ripe or nearly ripe using sickles. Therefore called "pre-domestication cultivation"

ANIMAL HUSBANDRY: Term refers to all types of animal rearing practices, ranging from chicken to cattle.

Pastoralism: Term normally used to refer to subsistence-oriented livestock production in which some animals or animal products are sold or bartered for food or other commodities, but family reproduction relies largely on the herds. Animals featured in this way of life vary according to regions and include cattle, sheep, goats, camels, horses.

Centers of early domestication

Southwest Asia

- Mobile Hunter-Gatherers

- Sedentary hunter-gatherers

- Sedentary farming communities

Credited with domesticating: Dog, pig, goat, sheep, wheat, barley, oat, peas, lentils, apples.

China

- Mobile Hunter-Gatherers

- Sedentary hunter-gatherers

- Sedentary farming communities

Credited with domesticating: Rice

Africa

Credited with domesticating: Sorghum, cattle

Mesoamerica

- Mobile hunter-gatherers
- Small mobile farming communities
- Sedentary farming communities

  Credited with domesticating: Squash, pumpkin, corn, sunflower

HUMAN VARIATION AND ADAPTATION

One of the notable characteristics of the human species today is its great variability. Human diversity has long fascinated people, but unfortunately it also has led to discrimination. In this chapter we will attempts to address the following questions:

- **What are the causes of physical variability in modern animals?**

- **Is the concept of race useful for studying human physical variation?**

- **Are there differences in intelligence from one population to another?**

Variation and evolution

Human genetic variation generally is distributed in such a continuous range, with varying clusters of frequency.

Ex. Our hair is curly or straight, our skin is lightly to heavily pigmented, and in height we range from short to tall.

The significance we give our variations, the way we perceive them (in fact, whether or not we perceive them at all) is determined by our culture.

Many behavioral traits are learned or acquired by living in a society; other characteristics, such as blue eyes, are passed on physically by heredity. Environment affects both.

The physical characteristics of both populations and individuals are a product of the interaction between genes and environments.

Ex. One's genes predispose one to a particular skin color, but the skin color one actually has is strongly affected by environmental factors such as the amount of solar radiation.

For most characteristics, there are within the gene pool of Homo sapiens variant forms of genes, known as alleles.

In the color of an eye, the shape of a hand, the texture of skin, many variations can occur.

This kind of variability, found in many animal species, signifies a rich potential for new combinations of characteristics in future generations. A species faced with changing environmental conditions has within its gene pool the possibility of producing individuals with traits appropriate to its altered life. Many may not achieve reproductive success, but those whose physical characteristics enable them to do well in the new environment will usually reproduce, so that their genes will become more common in subsequent generations. Thus, humankind has been able to occupy a variety of environments.

A major expansion into new environments was under way by the time *Homo erectus* appeared on the scene. Populations of this species were living in Africa, Southeast Asia, Europe and China. The differentiation of animal life is the result of selective pressures that, through the Pleistocene, differed from one region to another. Coupled with differing selective pressures were geographical features that restricted or prevented gene flow between populations of different faunal regions.

Ex. The conditions of life were quite different in China, which lies in the temperate zone, than they were in tropical Southeast Asia.

Genetic variants will be expressed in different frequencies in these geographically dispersed populations.

Ex. In the Old World, populations of Homo sapiens living in the tropics have a higher frequency of genes for dark skin than do those living in more northerly regions.

In blood type, H. sapiens shows four distinct groups (A, B, O or AB):

- The frequency of the O allele is highest in Native Americans, especially

  among some populations native to South America;

- The highest frequencies of the allele for Type A tend to be found among

  certain European populations;

- The highest frequencies of the B allele are found in some Asian populations.

The Meaning of Race

Early anthropologists uIcd to explore the nature of human species by systematically classifying H. sapiens into subspecies or races, based on geographic location and physical features such as skin color, body size, head shape and hair texture. Such classifications were continually challenged by the presence of individuals who did not fit the categories.

The fact is, generalized references to human types such as "Asiatic" or "Mongoloid", "European" or "Caucasoid", and "African" or "Negroid" were at best mere statistical abstractions about populations in which certain physical features appeared in higher frequencies than in other populations.

No example of "pure" racial types could be found.

These categories turned out to be neither definitive nor particularly helpful. The visible traits were found to occur not in abrupt shifts from population to population, but in a continuum that changed gradually. Also one trait might change gradually over a north-south gradient, while another might show a similar change from east to west.

Human skin color becomes progressively darker as one moves from northern Europe to central Africa, while blood type B becomes progressively more common as one moves from western to eastern Europe.

Race as a biological concept

To understand why the racial approach to human variation has been so unproductive, we must first understand the race concept in strictly biological terms.

In biology, a race is defined as a population of a species that differs in the frequency of different variants of some gene or genes from other populations of the same species. Three important things to note about this definition:

- it is arbitrary. There is no agreement on how many genetic differences it takes to make a race. For some, different frequencies in the variants of one gene are sufficient; for others, differences in frequencies involving several genes were necessary. The number of genes and precisely which ones are the more important for defining races are still open to debate;

- it does not mean that any race has exclusive possession of any particular variant of any gene or genes. In human terms, the frequency of the allele for blood group O may be high in one population and low in another, but it is present in both. Races are genetically "open", meaning that gene flow takes place between them. Thus one can easily see the fallacy of any attempt to identify "pure" races: if gene flow cannot take place between two populations, either directly or indirectly through intermediate populations, then they are not races, but are separate species;

- individuals of one race will not necessarily be distinguishable from those of another. In fact, the differences between individuals within a population are generally greater than the differences between populations.

The concept of human races

As a device for understanding physical variation in humans, the biological race concept has serious drawbacks:

- the category is arbitrary, which makes agreement on any given classification difficult, if not impossible. For example, if one researcher emphasizes skin color, while another emphasizes blood group differences, they will not classify people in the same way. What has happened is that human populations have grown in the course of human evolution, and with this growth have come increased opportunities for contact and gene flow between populations. Since the advent of food production, the process has accelerated as higher birth rates and periodic food shortages have prompted the movement of farmers from their homelands to other places;

- things are complicated even more because humans are so complicated genetically;

- "race" exists as a cultural, as well as a biological, category. In various ways, cultures define religious, linguistic and ethnic groups as races, thereby confusing linguistic and cultural traits with physical traits;

- to make the matter even worse, this confusion of social with biological factors is frequently combined with attitudes (racism) that are then taken as excuses to exclude whole categories of people from certain roles or positions in society. In the United States, for example, the idea of race originated in the 18th century to refer to the diverse peoples - European settlers, conquered Indians, and Africans imported as slaves - that were brought together in colonial North America. This racial worldview assigned some groups to perpetual low status on the basis of their supposedly biological inferiority, while access to privilege, power and wealth was reserved for favored groups of European descent.

There has been a lot of debate not just about how many human races there may be, but about what "race" is and is not. Often forgotten is the fact that a race, even if it can be defined biologically, is the result of the operation of evolutionary process. Because it is these processes rather than racial categories themselves in which we are really interested, most anthropologists have abandoned the race concept as being of no particular utility. Instead, they prefer to study the distribution and significance of specific, genetically based characteristics, or else the characteristics of small breeding populations that are, after all, the smallest units in which evolutionary change occurs.

Physical variables

Not only have attempts to classify people into races proven counterproductive, it has also become apparent that the amount of genetic variation in humans is relatively low, compared to that of other primate species.

Nonetheless, human biological variation is a fact of life, and physical anthropologists have learned a great deal about it. Much of it is related to climatic adaptation. A correlation has been noted between body build and climate:

Generally, people native to regions with cold climates tend to have greater body bulk (not to be equated with fat) relative to their extremities (arms and legs) than do people native to regions with hot climates, who tend to be long and slender. Interestingly, these differences show up as early as the time of Homo erectus.

Certain body builds are better suited to particular living conditions than others.

- People with larger body bulk and shorter extremities may suffer more from summer heat than someone whose extremities are long and whose body is slender. But they will conserve needed body heat under cold conditions. The reason is that a bulky body tends to conserve more heat than a less bulky one, since it has less surface relative to volume;

- People living in hot, open country, by contrast, benefit from a body build that can get rid of excess heat quickly so as to keep from overheating; for this, long extremities and a slender body, which increase surface area relative to volume, are advantageous.

Anthropologists have also studied such body features as nose, eye shape, hair textures and skin color in relation to climate.

Ex. Subject to tremendous variation, skin color is a function of four factors: transparency or thickness of the skin, distribution of blood vessels, and amount of carotene and melanin in a given area of skin. Exposure to sunlight increases the amount of melanin, darkening the skin.

Natural selection has favored heavily pigmented skin as protection against the strong solar radiation of equatorial latitudes. In northern latitudes, natural selection has favored relatively depigmented skins, which can utilize relatively weak solar radiation in the production of Vitamin D. Selective mating, as well as geographical location, plays a part in skin color distribution.

Continuing human biological evolution

In the course of their evolution, humans in all parts of the world came to rely on cultural rather than biological adaptation for their survival. Nevertheless, as they spread beyond their tropical homeland into other parts of the world, they did develop considerable physical variation from one population to another.

The forces responsible for this include:

- genetic drift, especially at the margins of their range where small populations were isolated for varying amounts of time;
- biological adaptation to differing climates.

Although much of this physical variation can still be seen in human populations today, the increasing effectiveness of cultural adaptation has often reduced its importance. Cultural practices today are affecting the human organism in important, often surprising, ways.

The probability of alterations in human biological makeup induced by culture raises a number of important questions. By trying to eliminate genetic variants, are we weakening the gene pool by allowing people with hereditary diseases and defects to reproduce? Are we reducing chances for genetic variation by trying to control population size? We are not sure of the answers to these questions.

GNU Free Documentation License

Version 1.2, November 2002

0. PREAMBLE

The purpose of this License is to make a manual, textbook, or other functional and useful document "free" in the sense of freedom: to assure everyone the effective freedom to copy and redistribute it, with or without modifying it, either commercially or noncommercially. Secondarily, this License preserves for the author and publisher a way to get credit for their work, while not being considered responsible for modifications made by others.

This License is a kind of "copyleft", which means that derivative works of the document must themselves be free in the same sense. It complements the GNU General Public License, which is a copyleft license designed for free software.

We have designed this License in order to use it for manuals for free software, because free software needs free documentation: a free program should come with manuals providing the same freedoms that the software does. But this License is not limited to software manuals; it can be used for any textual work, regardless of subject matter or whether it is published as a printed book. We recommend this License principally for works whose purpose is instruction or reference.

1. APPLICABILITY AND DEFINITIONS

This License applies to any manual or other work, in any medium, that contains a notice placed by the copyright holder saying it can be distributed under the terms of this License. Such a notice grants a world-wide, royalty-free license, unlimited in duration, to use that work under the conditions stated herein. The "Document", below, refers to any such manual or work. Any member of the public is a licensee, and is addressed as "you". You accept the license if you copy, modify or distribute the work in a way requiring permission under copyright law.

A "Modified Version" of the Document means any work containing the Document or a portion of it, either copied verbatim, or with modifications and/or translated into another language.

A "Secondary Section" is a named appendix or a front-matter section of the Document that deals exclusively with the relationship of the publishers or authors of the Document to the Document's overall subject (or to related matters) and contains nothing that could fall directly within that overall subject. (Thus, if the Document is in part a textbook of mathematics, a Secondary Section may not explain any mathematics.) The relationship could be a matter of historical connection with the subject or with related matters, or of legal, commercial, philosophical, ethical or political position regarding them.

The "Invariant Sections" are certain Secondary Sections whose titles are designated, as being those of Invariant Sections, in the notice that says that the Document is released under this License. If a section does not fit the above definition of Secondary then it is not allowed to be designated as Invariant. The Document may contain zero Invariant Sections. If the Document does not identify any Invariant Sections then there are none.

The "Cover Texts" are certain short passages of text that are listed, as Front-Cover Texts or Back-Cover Texts, in the notice that says that the Document is released under this License. A Front-Cover Text may be at most 5 words, and a Back-Cover Text may be at most 25 words.

A "Transparent" copy of the Document means a machine-readable copy, represented in a format whose specification is available to the general public, that is suitable for revising the document straightforwardly with generic text editors or (for images composed of pixels) generic paint programs or (for drawings) some widely available drawing editor, and that is suitable for input to text formatters or for automatic translation to a variety of formats suitable for input to text formatters. A copy made in an otherwise Transparent file format whose markup, or absence of markup, has been arranged to thwart or discourage subsequent modification by readers is not Transparent. An image format is not Transparent if used for any substantial amount of text. A copy that is not "Transparent" is called "Opaque".

Examples of suitable formats for Transparent copies include plain ASCII without markup, Texinfo input format, LaTeX input format, SGML or XML using a publicly available DTD, and standard-conforming simple HTML, PostScript or PDF designed for human modification. Examples of transparent image formats include PNG, XCF and JPG. Opaque formats include proprietary formats that can be read and edited only by proprietary word processors, SGML or XML for which the DTD and/or processing tools are not generally available, and the machine-generated HTML, PostScript or PDF produced by some word processors for output purposes only.

The "Title Page" means, for a printed book, the title page itself, plus such following pages as are needed to hold, legibly, the material this License requires to appear in the title page. For works in formats which do not have any title page as such, "Title Page" means the text near the most prominent appearance of the work's title, preceding the beginning of the body of the text.

A section "Entitled XYZ" means a named subunit of the Document whose title either is precisely XYZ or contains XYZ in parentheses following text that translates XYZ in another language. (Here XYZ stands for a specific section name mentioned below, such as "Acknowledgements", "Dedications", "Endorsements", or "History".) To "Preserve the Title" of such a section when you modify the Document means that it remains a section "Entitled XYZ" according to this definition.

The Document may include Warranty Disclaimers next to the notice which states that this License applies to the Document. These Warranty Disclaimers are considered to be included by reference in this License, but only as regards disclaiming warranties: any other implication that these Warranty Disclaimers may have is void and has no effect on the meaning of this License.

2. VERBATIM COPYING

You may copy and distribute the Document in any medium, either commercially or noncommercially, provided that this License, the copyright notices, and the license notice saying this License applies to the Document are reproduced in all copies, and that you add no other conditions whatsoever to those of this License. You may not use technical measures to obstruct or control the reading or further copying of the copies you make or distribute. However, you may accept compensation in exchange for copies. If you distribute a large enough number of copies you must also follow the conditions in section 3.

You may also lend copies, under the same conditions stated above, and you may publicly display copies.

3. COPYING IN QUANTITY

If you publish printed copies (or copies in media that commonly have printed covers) of the Document, numbering more than 100, and the Document's license notice requires Cover Texts, you must enclose the copies in covers that carry, clearly and legibly, all these Cover Texts: Front-Cover Texts on the front cover, and Back-Cover Texts on the back cover. Both covers must also clearly and legibly identify you as the publisher of these copies. The front cover must present the full title with all words of the title equally prominent and visible. You may add other material on the covers in addition. Copying with changes limited to the covers, as long as they preserve the title of the Document and satisfy these conditions, can be treated as verbatim copying in other respects.

If the required texts for either cover are too voluminous to fit legibly, you should put the first ones listed (as many as fit reasonably) on the actual cover, and continue the rest onto adjacent pages.

If you publish or distribute Opaque copies of the Document numbering more than 100, you must either include a machine-readable Transparent copy along with each Opaque copy, or state in or with each Opaque copy a computer-network location from which the general network-using public has access to download using public-standard network protocols a complete Transparent copy of the Document, free of added material. If you use the latter option, you must take reasonably prudent steps, when you begin distribution of Opaque copies in quantity, to ensure that this Transparent copy will remain thus accessible at the stated location until at least one year after the last time you distribute an Opaque copy (directly or through your agents or retailers) of that edition to the public.

It is requested, but not required, that you contact the authors of the Document well before redistributing any large number of copies, to give them a chance to provide you with an updated version of the Document.

4. MODIFICATIONS

You may copy and distribute a Modified Version of the Document under the conditions of sections 2 and 3 above, provided that you release the Modified Version under precisely this License, with the Modified Version filling the role of the Document, thus licensing distribution and modification of the Modified Version to whoever possesses a copy of it. In addition, you must do these things in the Modified Version:

A. Use in the Title Page (and on the covers, if any) a title distinct from that of the Document, and from those of previous versions (which should, if there were any, be listed in the History section of the Document). You may use the same title as a previous version if the original publisher of that version gives permission.

B. List on the Title Page, as authors, one or more persons or entities responsible for authorship of the modifications in the Modified Version, together with at least five of the principal authors of the Document (all of its principal authors, if it has fewer than five), unless they release you from this requirement.

C. State on the Title page the name of the publisher of the Modified Version, as the publisher.

D. Preserve all the copyright notices of the Document.

E. Add an appropriate copyright notice for your modifications adjacent to the other copyright notices.

F. Include, immediately after the copyright notices, a license notice giving the public permission to use the Modified Version under the terms of this License, in the form shown in the Addendum below.

G. Preserve in that license notice the full lists of Invariant Sections and required Cover Texts given in the Document's license notice.

124

H. Include an unaltered copy of this License.

I. Preserve the section Entitled "History", Preserve its Title, and add to it an item stating at least the title, year, new authors, and publisher of the Modified Version as given on the Title Page. If there is no section Entitled "History" in the Document, create one stating the title, year, authors, and publisher of the Document as given on its Title Page, then add an item describing the Modified Version as stated in the previous sentence.

J. Preserve the network location, if any, given in the Document for public access to a Transparent copy of the Document, and likewise the network locations given in the Document for previous versions it was based on. These may be placed in the "History" section. You may omit a network location for a work that was published at least four years before the Document itself, or if the original publisher of the version it refers to gives permission.

K. For any section Entitled "Acknowledgements" or "Dedications", Preserve the Title of the section, and preserve in the section all the substance and tone of each of the contributor acknowledgements and/or dedications given therein.

L. Preserve all the Invariant Sections of the Document, unaltered in their text and in their titles. Section numbers or the equivalent are not considered part of the section titles.

M. Delete any section Entitled "Endorsements". Such a section may not be included in the Modified Version.

N. Do not retitle any existing section to be Entitled "Endorsements" or to conflict in title with any Invariant Section.

O. Preserve any Warranty Disclaimers.

If the Modified Version includes new front-matter sections or appendices that qualify as Secondary Sections and contain no material copied from the Document, you may at your option designate some or all of these sections as invariant. To do this, add their titles to the list of Invariant Sections in the Modified Version's license notice. These titles must be distinct from any other section titles.

You may add a section Entitled "Endorsements", provided it contains nothing but endorsements of your Modified Version by various parties--for example, statements of peer review or that the text has been approved by an organization as the authoritative definition of a standard.

You may add a passage of up to five words as a Front-Cover Text, and a passage of up to 25 words as a Back-Cover Text, to the end of the list of Cover Texts in the Modified Version. Only one passage of Front-Cover Text and one of Back-Cover Text may be added by (or through arrangements made by) any one entity. If the Document already includes a cover text for the same cover, previously added by you or by arrangement made by the same entity you are acting on behalf of, you may not add another; but you may replace the old one, on explicit permission from the previous publisher that added the old one.

The author(s) and publisher(s) of the Document do not by this License give permission to use their names for publicity for or to assert or imply endorsement of any Modified Version.

5. COMBINING DOCUMENTS

You may combine the Document with other documents released under this License, under the terms defined in section 4 above for modified versions, provided that you include in the combination all of the Invariant Sections of all of the original documents, unmodified, and list them all as Invariant Sections of your combined work in its license notice, and that you preserve all their Warranty Disclaimers.

The combined work need only contain one copy of this License, and multiple identical Invariant Sections may be replaced with a single copy. If there are multiple Invariant Sections with the same name but different contents, make the title of each such section unique by adding at the end of it, in parentheses, the name of the original author or publisher of that section if known, or else a unique number. Make the same adjustment to the section titles in the list of Invariant Sections in the license notice of the combined work.

In the combination, you must combine any sections Entitled "History" in the various original documents, forming one section Entitled "History"; likewise combine any sections Entitled "Acknowledgements", and any sections Entitled "Dedications". You must delete all sections Entitled "Endorsements."

6. COLLECTIONS OF DOCUMENTS

You may make a collection consisting of the Document and other documents released under this License, and replace the individual copies of this License in the various documents with a single copy that is included in the collection, provided that you follow the rules of this License for verbatim copying of each of the documents in all other respects.

You may extract a single document from such a collection, and distribute it individually under this License, provided you insert a copy of this License into the extracted document, and follow this License in all other respects regarding verbatim copying of that document.

7. AGGREGATION WITH INDEPENDENT WORKS

A compilation of the Document or its derivatives with other separate and independent documents or works, in or on a volume of a storage or distribution medium, is called an "aggregate" if the copyright resulting from the compilation is not used to limit the legal rights of the compilation's users beyond what the individual works permit. When the Document is included in an aggregate, this License does not apply to the other works in the aggregate which are not themselves derivative works of the Document.

If the Cover Text requirement of section 3 is applicable to these copies of the Document, then if the Document is less than one half of the entire aggregate, the Document's Cover Texts may be placed on covers that bracket the Document within the aggregate, or the electronic equivalent of covers if the Document is in electronic form. Otherwise they must appear on printed covers that bracket the whole aggregate.

8. TRANSLATION

Translation is considered a kind of modification, so you may distribute translations of the Document under the terms of section 4. Replacing Invariant Sections with translations requires special permission from their copyright holders, but you may include translations of some or all Invariant Sections in addition to the original versions of these Invariant Sections. You may include a translation of this License, and all the license notices in the Document, and any Warranty Disclaimers, provided that you also include the original English version of this License and the original versions of those notices and disclaimers. In case of a disagreement between the translation and the original version of this License or a notice or disclaimer, the original version will prevail.

If a section in the Document is Entitled "Acknowledgements", "Dedications", or "History", the requirement (section 4) to Preserve its Title (section 1) will typically require changing the actual title.

9. TERMINATION

You may not copy, modify, sublicense, or distribute the Document except as expressly provided for under this License. Any other attempt to copy, modify, sublicense or distribute the Document is void, and will automatically terminate your rights under this License. However, parties who have received copies, or rights, from you under this License will not have their licenses terminated so long as such parties remain in full compliance.

10. FUTURE REVISIONS OF THIS LICENSE

The Free Software Foundation may publish new, revised versions of the GNU Free Documentation License from time to time. Such new versions will be similar in spirit to the present version, but may differ in detail to address new problems or concerns. See http://www.gnu.org/copyleft/.

Each version of the License is given a distinguishing version number. If the Document specifies that a particular numbered version of this License "or any later version" applies to it, you have the option of following the terms and conditions either of that specified version or of any later version that has been published (not as a draft) by the Free Software Foundation. If the Document does not specify a version number of this License, you may choose any version ever published (not as a draft) by the Free Software Foundation

External links

- GNU Free Documentation License (Wikipedia article on the license)

- Official GNU FDL webpage

www.ingramcontent.com/pod-product-compliance
Lightning Source LLC
Chambersburg PA
CBHW021432180326
41458CB00001B/243